DETERMINISTIC METHODS IN SYSTEMS HYDROLOGY

IHE DELFT LECTURE NOTE SERIES

Deterministic Methods in Systems Hydrology

JAMES C.I. DOOGE
J. PHILIP O'KANE

CRC Press

Taylor & Francis Group
Boca Raton London New York

CRC Press is an imprint of the
Taylor & Francis Group, an **informa** business

CRC Press
Taylor & Francis Group
6000 Broken Sound Parkway NW, Suite 300
Boca Raton, FL 33487-2742

Visit the Taylor & Francis Web site at
http://www.taylorandfrancis.com

and the CRC Press Web site at
http://www.crcpress.com

To the memory of Eamonn Nash

Table of Contents

List of Figures

List of Tables

Preface

This work is intended to survey the basic theory that underlies the multitude of parameter-rich models that dominate the hydrological literature today. It is concerned with the application of the equation of continuity (which is the fundamental theorem of hydrology) in its complete form combined with a simplified representation of the principle of conservation of momentum. Since the equation of continuity can be expressed in linear form by a suitable choice of state variables and is also parameter-free, it can be readily formulated at all scales of interest. In the case of the momentum equation, the inherent non-linearity results in problems of parameter specification at each particular scale of interest.

The approach is that of starting with a simplified but rigorous analysis in order to gain insight into the essential characteristics of the system operation and then using this insight to decide which restrictive simplification to relax in the next phase of the analysis. The benefits of this approach have been well expressed by Pedlosky (1987)[1]

> "One of the key features of geophysical fluid dynamics is the need to combine approximate forms of the basic fluid-dynamical equations of motion with careful and precise analysis. The approximations are required to make any progress possible, while precision is demanded to make the progress meaningful."

The replacement of empirical correlation analysis by complex parameter-rich models represents an improvement in the matching of predictive schemes to individual known data sets but does not advance our basic knowledge of hydrological processes firmly based on hydrologic theory.

The original version of the text was prepared at the invitation of Professor Mostertman some twenty-five years ago for the benefit of international postgraduate students at UNESCO-IHE Delft and has been used as a basis for lectures in subsequent years. It deals with the basic principles of some important deterministic methods in the systems approach to problems in hydrology. As such, it reflects the classical period of development in the application of systems theory to hydrology. In these lectures attention was

---

[1] Paragraph 2 in Joseph Pedlosky's "Preface to the First Edition" in his book *"Geophysical Fluid Dynamics"*, second edition, Springer-Verlag, pp. vii and viii.

confined to deterministic inputs as the methods appropriate to stochastic inputs were dealt with elsewhere.

The objectives of the course of lectures on "Deterministic Methods in Systems Hydrology" were

1. To introduce the elements of systems science as applied to hydrologic problems in such a way that students can appreciate the nature of the approach and can, if they wish, extend their knowledge of it by reading the relevant literature;
2. To approach flood prediction and the hydrologic methods of flood routing as problems in linear systems theory, so as to clarify the basic assumptions inherent in these methods, to extend the scope of these classical methods, and to evaluate their accuracy;
3. To review and evaluate some deterministic models of components of the hydrologic cycle, with a view to assembling the most appropriate model of catchment response, for a particular problem in applied hydrology.

The material is developed in two parts. The four chapters in the first part present the systems viewpoint, the nature of hydrologic systems, some systems mathematics and their application to the black-box analysis of direct storm runoff. Four additional chapters form the second part and cover linear conceptual models of direct runoff, the fitting of conceptual models to data, simple models of subsurface flow, and non-linear deterministic models. A set of exercises completes the exposition of the material.

It was not anticipated that the student would be able as a result of these lectures to master the complexities of the theory and all the details of individual models. Rather it was hoped that he or she would gain a general appreciation of the systems approach to hydrologic problems. Such an appreciation could serve as a foundation for a more complete understanding of the details in this text and in the cited references.

The original version of the text has been extensively edited. New material has also been added: the equivalence theorem of linear cascades in series and parallel, and the limiting cases of cascades, with and without lateral inflow, as seen in shape factor diagrams. Four new appendices present additional material extending the treatment of various topics.

Appendix B shows that de-convolution of linear systems, and by extension the inversion of non-linear systems, is in general an ill-posed problem. Imposing mass conservation is not sufficient to ensure that the problem is well-posed. Additional assumptions are required. To this end, we include in appendix A, a detailed description of the computer program PICOMO, which is referred to extensively in the text. It contains approximately twenty linear conceptual models built using various assumptions on lateral inflow, translation in space, and storage delay. These all lead to well-posed problems of system identification. The reader is encouraged to experiment with the program, which can be downloaded through the IHE website: http://www.ihe.nl/. The reader may wish to compare or combine PICOMO with other more recent hydrological toolboxes, which can be requested by e-mail from http://ewre.cv.ic.ac.uk/software/toolkit.htm, or http://www.nuigalway.ie/hydrology/.

Linear methods of analysis require a clear understanding of the nature and occurrence of strong non-linearities in the relevant processes. Appendices C and D address these questions. Appendix C presents the non-linear theory of isothermal movement of liquid water and water vapour through the unsaturated zone. In the case of bare soil at the scale of one meter, two pairs of non-linearities present themselves as switches in the surface boundary conditions of the governing partial differential equation. The outer pair represents alternating wet and dry periods when the atmosphere switches the surface flux of water either into or out of the soil. The inner pair represents the intermittent switching to soil control of the surface flux. Appendix D discusses the linearisation of the non-linear equations of open channel flow, their solution as a problem in linear systems theory, and the errors of linearisation.

The cited references have also been supplemented to cover subsequent developments in the topics dealt with in the original text and in the new appendices. These are not intended to provide a comprehensive review of current literature but rather to highlight key publications that deal with significant extensions of the material.

CHAPTER 1

The Systems Viewpoint

1.1 NATURE OF SYSTEMS APPROACH

System

Before commencing our discussion of deterministic methods in hydro-logic systems it is necessary to be clear as to what we mean by a *system*. The word is much used nowadays both in scientific and non-scientific writing. Even if we confine ourselves to the scientific literature, there is a bewildering range of definitions of what is meant by a system and what is meant by the systems approach. For the purposes of our present discussion, a system may be defined as (Dooge, 1968, p. 58)

> Any structure, device, scheme, or procedure,
> real or abstract,
> that inter-relates
> in a given time reference
> an input, cause or stimulus,
> of matter, energy or information
> and
> an output, effect or response
> of information, energy or matter.

The first line of the above definition emphasises that anything at all that consists of *connected parts* may be considered as a system. The second line emphasises that a system may be real as in the case of an actual

Model

catchment area or abstract as in the case of a *model* of the operation of that catchment based on digital simulation.

The following lines emphasise the important characteristics of dynamic systems that link *input* and *output*. The latter terminology is that usually used by the engineer but the physicist would tend to refer to them as *cause* and *effect* and a biologist as *stimulus* and *response*. The terms are used as alternatives in the above definition in order to stress that, from the systems viewpoint, there is no basic difference between the par-ticular systems studied by the engineer, the physicist and the biologist. In the definition, the input and the output may consist of matter, or energy, or information. These alternative categories are listed in different order for the input and the output respectively, to emphasise that input and output need not belong to the same category. In the case of *dynamic systems*, input and output must be defined in terms of a given time reference.

The *scale of the time reference* in which a dynamic system is considered, may vary according to the particular aspect of the system under study. Thus, in the case of hydrologic catchments, the flood producing power of the catchment may be considered in a time reference of hours or days. On the other hand, consideration of the sustained baseflow may require a time scale of months and possibly years. Consideration of the development of the drainage network from a geomorphological viewpoint, may involve a time reference expressed in centuries.

In applied science, our concern is predicting the output from the system of interest. This problem can be approached from a number of points of view. Firstly, one can adopt what might be called the *mathematical physics approach*, which seeks (a) to establish differential equations governing the physical phenomena involved, (b) to formulate the set of equations and boundary conditions for the particular system under study, and (c) to solve the resulting problem for a given input. An example would be the use of the St. Venant equation, to solve problems of overland flow or unsteady flow in a channel network. Freeze (1972) provides further discussion with particular reference to subsurface flow.

Mathematical physics approach

A second and sharply contrasting approach is that known as *black-box analysis*. In this approach, we do not use (at least initially) our knowledge of either (a) the physics of the processes involved, nor (b) the exact nature of the system. Instead, an attempt is made to extract from past records of input–output events on the system under examination, enough knowledge of the operation of that particular system, to serve as the basis for predicting its output due to other specified inputs. Dooge (1973, pp. 75–101) has reviewed the development of this approach, as it relates to the classical methods of applied hydrology.

Black-box analysis

Between these two extremes, lies the approach based on what are termed *conceptual models*. Even though both extremes represent "conceptual models" in the broad sense of that phrase, the term is usually limited in systems hydrology to a particular type of model. Conceptual models in hydrology consist of simple arrangements of elements whose structure and parameter values are chosen to simulate the behaviour of the hydrologic system under study. Such conceptual models may take the form of closed systems or open systems depending on the nature of the prototype (Bertalanffy, 1968, pp. 39–41). It is important to include (a) feedbacks between the components of the model (Bertalanffy, 1968, pp. 42–44), and (b) thresholds, i.e. points of concentrated non-linearity, which may switch the control of the system operation from one component to another. Rather than three isolated classes of model — black-box, conceptual models, equations of mathematical physics — there is in fact a complete spectrum of models ranging from pure black-box analysis, which makes almost no physical assumptions, to a highly complex approach based on equations derived from continuum mechanics.

Conceptual models

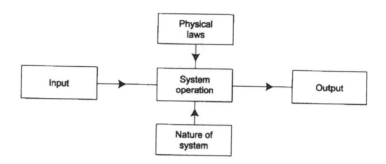

Figure 1.1. The concept of system operation.

The concept of *system* operation is illustrated in Figure 1.1 (Dooge, 1968, p. 59). In the classical approach based on mathematical physics, our knowledge of the physical laws involved and the information available to us concerning the nature of the system, enables us to describe the system operation in terms of a set of mathematical equations. In the case of a complicated and non-homogeneous hydrological system, this will give rise to a large set of complicated partial differential equations. A solution for this set of equations corresponding to the input of interest is then sought, but may prove difficult to obtain even by numerical methods using a large computer.

In the black-box approach, the vertical relationships in Figure 1.1 are ignored, and an attempt is made to obtain a mathematical characterisation of the operation of the particular system under study, from a past record of input–output data. If this can be done, and, if it can be used as a reliable basis for predicting the output due to other specified input, the problem has been solved in terms of the horizontal elements in Figure 1.1. In the *Systems approach* systems approach, we are concerned primarily with the conversion of input to output, and we ignore (at least initially) the components of the system, their connection with one another, and the physical laws involved.

While the development of a consistent theory of black-box analysis is a relatively recent development in hydrology, the use of such an approach as a tool in applied hydrology is not. Long before the theory was developed and applied, hydrologists in the field were using *unit hydrograph* methods for flood prediction and using simple conceptual models for flood routing. By developing a consistent theory it has been possible to clarify the scope and limitation of the methods thus used on an empirical basis.

1.2 SYSTEMS TERMINOLOGY

As in every other discipline, a special terminology has grown up in connection with system studies. The meaning of the more important concepts and terms must be appreciated if any progress is to be made on the study of the subject or in the reading of the literature. A system has already been defined in the last section. A *complex system* may be divided into *sub-systems*, each of which can be identified as having a distinct input–output linkage. A sub-system may be further divided into sub-sub-systems. A sub-system, or a sub-sub-system not further sub-divided, is frequently referred to as a *component* of the system.

The mathematical function that characterises the manner in which a system converts input into output is frequently referred to as the *transfer function* of the system. Instead of describing a system in terms of its input, output and transfer function, the system is sometimes characterised by relationships between the input, the output and *the state of the system*.

The concept of *state* is a general one and the formulation of a system model in the terms of state variables is particularly useful in the case of non-linear systems. Any change in any variable of the system will produce a change of state. If a set of state variables is completely known, the state of the system is known at that instant. The change in state from instant to instant can be determined from the initial state, the evolutionary equations of the state variables, and the input. A further equation relating the output to the state in any given instant is required. The set of state variables defining the operation of a given system is not necessarily unique.

Linear system

The distinction between *linear* and *non-linear* is of vital importance in systems theory as it is in classical mechanics. The *analysis* and *synthesis* of linear systems can draw on the immense storehouse of linear mathematics for its tools and techniques. The special properties of linear systems will be dealt with in later Chapters. For the moment, it suffices to say that a linear system is one that has the property of *superposition in time*, and a non-linear system is one that does not have this property.

Time-invariant system

Another important distinction is between *time-invariant* and *time-variant* systems. A time-invariant system is one whose input–output relationship or input–output-state relationship does not depend on the time at which the input is applied. Most hydrologic systems are time-variant due to diurnal and seasonal variations. Nevertheless, the advantages of assuming hydrologic systems to be time-invariant are sufficiently attractive that these real variations in time are often neglected in practice, usually with impunity.

Deterministic system

The terms deterministic and stochastic recur frequently in the discussion of systems behaviour. In a *deterministic system*, identical inputs will always produce identical outputs. A *stochastic system* is a system which contains one or more components in which the relationship between the input and output is random rather than deterministic. Care must be taken to distinguish between the stochastic operation of a system, and the stochastic nature of an input to a system, whose operation on that input is deterministic. It is clear from the title of the present book that it is concerned with deterministic methods and therefore with the response of deterministic systems to deterministic inputs. A system in which an output cannot occur earlier than the corresponding input (i.e. the effect cannot precede the cause) is termed a *causal* or a *non-anticipative* system. The systems dealt with in hydrology are always treated as causal systems.

Causal system

It is necessary to distinguish between *continuous* and *discrete* systems and between continuous and discrete inputs and outputs of systems. A system is said to be continuous when the operation of the system takes place continuously and is said to be discrete when it changes its state only at discrete intervals of time. An input or an output of a system is said to be continuous, when its value is either known continuously, or can be sampled so frequently, that the record may be assumed to be continuous without appreciable error. An input or output is said to be discrete, if the value is only known or can only be sampled at finite time intervals. An input or an output is said to be *quantized* when the value changes at discrete intervals of time and holds a constant value between these intervals. Many records of rainfall, which are known only in terms of the volume during specified intervals of time, are in fact quantized records. Whereas hydrologic systems are continuous in their operation, the inputs and outputs may be available in either continuous, discrete or quantized form.

Lumped or distributed

The input and output variables, and the parameters of a system, may be *lumped* or *distributed*. A lumped variable or lumped parameter does not vary in any space dimension. In this case, spatial variation has been averaged out, or ignored because it is negligible. Thus the average rainfall over a catchment, when used as the input to a hydrological model, is a lumped input. Where the variation in one or more space dimensions is taken into account, the parameter is said to be distributed. The behaviour of lumped systems is governed by ordinary differential equations with time as the single independent variable. The behaviour of distributed systems is governed by partial differential equations with time and one or more spatial variables as the independent variables.

Memory of the system

The length of time during which the input affects the present state is referred to as the *memory of the system*. If a system has a zero memory, its state and its present output depend only on the present input. If a system has an infinite memory, the state and the output will depend on the whole past history of the system. In a system with a finite memory, its state and its output depend only on the history of the system for a previous length of time equal to the memory.

A *stable system* is one in which, when the input is bounded, the output is similarly bounded. In hydrology, almost all our systems are stable, and most of them highly stable. Generally speaking, when the input to a hydrologic system is bounded, the bound on the output is considerably less than that of the input. Stability in a system is promoted by the presence

Negative feedback loops

in it of *negative feedback loops*. Some systems are stable against small changes in the environment, but not stable in respect of large changes. Other systems are highly *adaptive* and can learn from past history how to optimise their behaviour according to some criterion.

In the pure *black-box approach*, attention is focused on the interrelationship between input and output without any reference to the physical processes involved in the transformation or in the nature of the actual system. This general model may be represented by

$$y(t) = H[x(t)] \tag{1.1}$$

where $x(t)$ represents the input to the system, H represents the operation of the system on that input and $y(t)$ represents the resulting output. There are three basic problems in black-box analysis as shown in Table 1.1.

Prediction

Firstly, there is the relatively straightforward *problem of prediction*, where the input $x(t)$ and the nature of the operator H are both known, and have to be combined to predict the output $y(t)$. Secondly, there is the *problem of system identification*, in which the input $x(t)$ and the output $y(t)$ are known, and it is required to find, and if possible to express mathematically, the nature of the system operation represented by H in equation (1.1). Finally, there is the *problem of signal detection*, in which the system operation H and the output $y(t)$ are both known, and it is required to deduce the nature of the input $x(t)$. It will be appreciated that the latter two problems are *inverse problems* and consequently likely to be more difficult than the direct problem of output prediction.

Identification

Detection

In contrast to *systems analysis* (which includes system identification, input detection, and output prediction) there are the techniques of system simulation. In *systems simulation* an attempt is made to design a system (real or abstract), which will simulate the conversion of input to output by the prototype system within the limits of accuracy required by the problem under study. Real *simulation systems* include physical models and special purpose analogs. Abstract models include both simple conceptual models and highly complex mathematical formulations of physical theory. In all simulations there is the problem of determining the optimum values for the parameters to be used in the model.

Simulation

Table 1.1. Classification of basic problems in systems analysis and synthesis.

Category	Type of problem	Input function $x(t)$	System operation $H[.]$	Output function $y(t)$
System analysis	1. Prediction	Given	Given	Unknown
	2. Identification	Given	Unknown	Given
	3. Detection	Unknown	Given	Given
System synthesis	Simulation	Given	Conceptual model Model structure Parameter values	Given

1.3 LINEAR TIME-INVARIANT SYSTEMS

As in all approaches in applied analysis, the first step is to make certain *simplifying assumptions*, which facilitate the solution of the problem, and then to evaluate the adequacy of these simple methods. Even if the simplified versions are inadequate, they frequently point the way towards the solution of the more complex problem. While $x(t)$ and $y(t)$ in equation (1.1) could be considered as vectors and thus may represent *multiple inputs* and *multiple outputs*, it is more convenient in a first discussion to treat them as single variables i.e. as *lumped inputs and outputs*. Assuming the inputs and outputs to be lumped in the first instance, corresponds to selecting models based on ordinary differential equations, before considering models based on the partial differential equations of mathematical physics.

Linear system

The next obvious simplification is to limit our attention, for the time being, to *linear systems*, and thus to take advantage of the great power of linear methods in mathematical analysis.

The linearity in the systems sense, as defined above, must be clearly distinguished from a relationship between x and y which is linear in the sense of the algebraic equation

$$y = a + bx \tag{1.4}$$

It should be noted that if an input–output pair $(\underline{x}, \underline{y})$ obeys the relationship given by equation (1.4), the pair $(a\underline{x}, a\underline{y})$ which would hold for a linear system in accordance with equations (1.2) and (1.3) (Table 1.2) does not satisfy the relationship given by equation (1.4).

Similarly it is necessary to distinguish between the superposition property of a linear system and the linearity of the regression equation

$$y = a_0 + a_1 x + a_2 x^2 \tag{1.5}$$

in which we have a linearity in the coefficients rather than in the variables (Clarke, 1973).

Table 1.2. Definition of linearity.

A system is said to be linear, if and only if, the response of a system to an input $\underline{x}(t)$, which is a linear combination of a number of elementary inputs $x_i(t)$

$$\underline{x}(t) = \sum_i c_i\, x_i(t) \tag{1.2}$$

is given by the output $\underline{y}(t)$, which is the same combination

$$\underline{y}(t) = \sum_i c_i\, y_i(t) \tag{1.3}$$

of the elementary outputs $y_i(t)$ corresponding to the respective elementary inputs $x_i(t)$.

The assumption of linearity allows us to write equation (1.1) in more explicit form. This can be done most conveniently by the use of the concept of an *impulse function* or *Dirac delta function*. Such a pseudo-function, or distribution (Schwartz, 1951), $\delta(t)$, is usually visualised as a limiting form of a pulse of some particular shape, as the duration of the pulse goes to zero at $t = 0$, while maintaining a constant area usually taken as unity (Aseltine, 1958, pp. 22–31). Thus a *unit impulse function* located at $t = \tau$ has the properties

Unit impulse function

$$\delta(t - \tau) = 0 \quad \text{when } t \neq \tau \tag{1.6a}$$

$$\int_{-\infty}^{\infty} \delta(t - \tau)\, \mathrm{d}\tau = 1 \tag{1.6b}$$

The mathematically more correct definition of a delta function is in terms of its ability to sift out specific values of a function $x(t)$,

$$x(t) = \int_{-\infty}^{\infty} x(\tau)\, \delta(t - \tau)\, \mathrm{d}\tau \tag{1.7}$$

If $x(t) = 1$, we recover (1.6b). The *impulse response* of a system, $h(t, \tau)$, is defined as the output from the system when the input takes the form of an impulse or delta function at the time $t = \tau$ i.e. if $x(t) = \delta(t - \tau)$, then $y(t) = h(t, \tau)$.

From the definition of the impulse response and the fact that equation (1.7) is a special form of equation (1.2), we get the corresponding form of equation (1.3) as

$$y(t) = \int_{-\infty}^{\infty} x(\tau)\, h(t, \tau)\, \mathrm{d}\tau \tag{1.8}$$

Table 1.3 presents the details of the derivation. It should be read initially column by column, from left to right.

It is clear from equation (1.8) that we can predict the output for any given input, by multiplication and integration, whenever the impulse response function is known. However, the problem of system identification i.e. of determining the impulse response function from given records of input and output, is seen to involve the solution of an integral equation. Hence, the system operator $H[x(t)]$ for lumped linear *time-variant*

Table 1.3. The integral equation for linear systems.

For a given point in time t	when the input is ...	the corresponding output is ...
Place a δ-fn at a second point in time τ	$\delta(t - \tau)$	$h(t, \tau)$
Scale the δ-fn with a slice of x at τ	$x(\tau)\, \mathrm{d}\tau\, \delta(t - \tau)$	$x(\tau)\, \mathrm{d}\tau\, h(t, \tau)$
Superimpose the slices by integration on τ	$x(t) = \int_{-\infty}^{\infty} x(\tau)\, \delta(t - \tau)\, \mathrm{d}\tau$	$y(t) = \int_{-\infty}^{\infty} x(\tau)\, h(t, \tau)\, \mathrm{d}\tau$

Conservation law

Heavily damped systems

systems, is a weighted integral of the input function $x(t)$, with a weighting function $h(t, \tau)$ which itself varies with time t. If the system obeys a *conservation law*, total input must equal total output. If we take the special case of $x(t) = y(t) = a$ *constant input/output rate per unit time* for all values of t, conservation is obviously satisfied, and we see from equation (1.8) that the integral of $h(t, \tau)$ with respect to τ must equal 1 for all values of time t. Integration of (1.8) with respect to t, and changing the order of integration on the right-hand side, generalises this result for all input/output pairs satisfying a conservation law. If all input/output pairs are non-negative, $h(t, \tau)$ must also be non-negative. *Heavily damped systems* have non-negative $h(t, \tau)$ and no oscillations.

If the system is assumed to be *time-invariant*, the impulse response at any time t will depend only on the time elapsed since the occurrence of the input. Hence

$$h(t, \tau) = h(t - \tau) \tag{1.9}$$

and the relationship between input and output may be written as

$$y(t) = \int_{-\infty}^{\infty} x(\tau)\, h(t - \tau)\, \mathrm{d}\tau \tag{1.10a}$$

The right-hand side is the well-known mathematical operation of *convolution*, which is usually written as

$$y(t) = x(t) * h(t) \tag{1.10b}$$

Convolution integral

The convolution, or folding (or *Faltung*) integral, is the integral of the product of two functions, x and h, the sum of whose arguments, τ and $t - \tau$, is a parameter t of the integral. Four operations take place in forming the integral: shift, fold, multiply, and integrate. In relation (1.10a), for a given value of t, the argument of $h(.)$ is replaced with $(t - \tau) = -(\tau - t)$. The new argument $(\tau - t)$ shifts the graph of the h function by t along the τ axis. The minus sign in front of the argument $-(\tau - t)$ folds the shifted function back along the τ axis i.e. reflects it in the vertical line at $\tau = t$. The shifted and folded graph of $h(.)$ on the τ axis, and the graph of the $x(\tau)$ function, are multiplied together for all values of τ. This product function is then integrated with respect to τ to give a measure of area, which is assigned to y at the given value of t, the shift. Repeating these four operations for all values of the shift t, defines the function $y(t)$. Shifting and folding are time-invariant operations when there is no change in the h function; multiplication and integration are linear operations.

The first attempt at developing methods of systems analysis will be concerned with equation (1.10) and hence with *lumped linear time-invariant systems*. In the classical approach based on mathematical physics, the equivalent problem to the solution of equation (1.10) is the solution of a set of ordinary linear differential equations with constant coefficients.

If the system to which we are applying equation (1.10) is *causal* the impulse response will be zero for all negative values of the argument. Consequently, there will be no contribution to the integral on the right hand side of the equation for values of τ greater than t, and the infinite upper limit in the integration can be replaced by t thus giving

$$y(t) = \int_{-\infty}^{t} x(\tau)\, h(t - \tau)\, d\tau \qquad (1.11)$$

Finite memory

If the systems has a *finite memory* M, then the impulse response system will be zero for all arguments greater than M. In this case there will be no contribution to the integral for values of τ less than $t-M$, so that the lower limit of the integral can be written as

$$y(t) = \int_{t-M}^{t} x(\tau)\, h(t - \tau)\, d\tau \qquad (1.12)$$

If the input is an *isolated* one i.e. if the time between successive inputs exceeds the memory of the system then $x(t)$ and $y(t)$ will be zero for negative arguments so that equation (1.11) can be written in the form

$$y(t) = \int_{0}^{t} x(\tau)\, h(t - \tau)\, d\tau \qquad (1.13)$$

Equations (1.12) and (1.13) can be combined by writing

$$y(t) = \int_{[t-M]}^{t} x(\tau)\, h(t - \tau)\, d\tau \qquad (1.14)$$

where $[t - M] = max(0, t - M)$ has the value zero for $t < M$, and the value $t-M$ for $t > M$.

In the present text, equation (1.13) will normally be used to describe a linear time-invariant system unless the circumstances of the problem indicate one of the other forms would be more suitable. Equation (1.10b) makes no distinction between the variables on the right hand side of the equation and it can be shown by a change of variable that equation (1.13) may be written as

$$y(t) = \int_{0}^{t} x(t - \tau)\, h(\tau)\, d\tau \qquad (1.15a)$$

Commutative

Associative

The latter form is sometimes more convenient to use. This establishes that the order is unimportant i.e. convolution is a *commutative* operation: $y = x * h = h * x$. It is also an *associative* operation: $y = (x * h_1) * h_2 = x * (h_1 * h_2)$. Commuting the order, we also find $y = x * (h_2 * h_1) = (x * h_2) * h_1$. Hence, if x is the input to two *sub-systems in series* with individual response functions h_1 and h_2 respectively, their overall response

Distributive

function is $h_1 * h_2$ and the order in which they are placed in the series does not matter. Finally, convolution is a *distributive* operation: $y = (x_1 + x_2) * h = x_1 * h + x_2 * h$ with an obvious interpretation as two identical *sub-systems in parallel*, one with x_1 as input, the other with x_2 as input.

These properties can be proved directly, or indirectly using the Laplace Transform. The Laplace Transform transforms the x, y and h functions above, from the t-space of their arguments, to the s-space of this linear transform. It is shown in Chapter 3 that convolution in t-space becomes multiplication in s-space. Since multiplication is also commutative, associative and distributive, convolution defines a "very reasonable multiplication on the real vector space of all piece-wise continuous functions of exponential order under the usual definitions of addition and scalar multiplication". See Kreider et al. (1966, pp. 183, 206, and 208).

Cascades of sub-systems

The most important application of the commutative, associative and distributive properties of convolution is to cascades of sub-systems in series and in parallel. We begin with a system, S_s, consisting of a cascade of n sub-systems in series, where the output from one sub-system is the input to the next. Each sub-system in the cascade also receives a lateral input, which is an arbitrary fraction of a single overall input to the cascade.

Let $h_i(t)$ be the impulse response of sub-system i. It receives as input, the output from sub-system $i - 1$ above it in the cascade, together with a lateral input equal to a fraction a_i of the overall input $x(t)$. The output from sub-system i in S_s by definition is

$$y_i(t) = h_i(t) * [a_i \cdot x(t) + y_{i-1}(t)], \quad y_0 = 0 \qquad i = 1, \ldots, n \qquad (1.15b)$$

where the weights are non-negative real numbers that sum to one:

$$a_i \geq 0, \quad \sum_1^n a_i = 1 \qquad\qquad (1.15c)$$

Iterating this recurrence relation gives

$$y_1(t) = h_1(t) * [a_1 \cdot x(t)]$$
$$y_2(t) = h_2(t) * [a_2 \cdot x(t) + y_1(t)]$$

$$\vdots \qquad\qquad (1.15d)$$

$$y_{n-1}(t) = h_{n-1}(t) * [a_{n-1} \cdot x(t) + y_{n-2}(t)]$$
$$y_n(t) = h_n(t) * [a_n \cdot x(t) + y_{n-1}(t)]$$

Substituting from top to bottom in the cascade, eliminating all intermediate outputs, we find the following nested expression for the output from the system S_s

$$y_n(t) = h_n(t) * [a_n \cdot x(t) + h_{n-1}(t) * [a_{n-1} \cdot x(t)$$
$$+ h_{n-2}(t) * [\cdots + h_2(t) * [a_2 \cdot x(t) + h_1(t) * [a_1 \cdot x(t)]] \cdots]]]$$

(1.15e)

Since convolution is associative, commutative and distributive this can be expanded as

$$y_n(t) = a_n \cdot x(t) * h_n(t)$$
$$+ a_{n-1} \cdot x(t) * h_n(t) * h_{n-1}(t)$$
$$\cdots$$
$$+ a_2 \cdot x(t) * h_n(t) * h_{n-1}(t) * \cdots * h_2(t)$$
$$+ a_1 \cdot x(t) * h_n(t) * h_{n-1}(t) * \cdots * h_2(t) * h_1(t)$$

(1.15f)

or, in more compact notation,

$$y_n(t) = x(t) * \left[\sum_{j=1}^{n} a_j \cdot g_j(t) \right]$$

(1.15g)

where the index j runs over n subsets of the cascade, starting with all n sub-systems in S_s ($j = 1$), and ending with the last sub-system in S_s on its own ($j = n$):

$$g_1(t) = h_1(t) * h_2(t) * \quad \cdots \cdots \cdots \quad * h_{n-1}(t) * h_n(t)$$
$$g_2(t) = \quad h_2(t) * \quad \cdots \cdots \cdots \quad * h_{n-1}(t) * h_n(t)$$
$$\vdots$$
$$g_j(t) = \quad h_j(t) * h_{j+1}(t) * \cdots * h_{n-1}(t) * h_n(t) \qquad (1.15h)$$
$$\vdots$$
$$g_{n-1}(t) = \quad h_{n-1}(t) * h_n(t)$$
$$g_n(t) = \quad h_n(t)$$

Sub-systems in parallel

These conclusions can be interpreted in terms of a new system, S_p, consisting of n *sub-systems in parallel*. Each parallel sub-system in S_p receives a fraction a_j of the overall input $x(t)$ and its output contributes directly to the overall system output $y_n(t)$. The jth parallel sub-system in S_p, can be described as (a) a sub-cascade of $(n - j + 1)$ adjacent sub-systems taken from the bottom end of the cascade S_s, or (b) a single equivalent sub-system with impulse response $g_j(t)$. Note that S_s and S_p are identical, when $a_1 = 1$, and $a_j = 0$ for $j = 2, \ldots, n$.

The overall output $y_n(t)$ from the two systems, S_s and S_p, will be equal, when the impulse response $g_j(t)$ of each parallel sub-system in S_p, is equal to the impulse response of the corresponding partial cascade running from j to n in S_s. This is the *Equivalence Theorem* for linear time-invariant sub-systems in series and parallel. The reader should draw the system diagrams for S_s and S_p.

Equivalence theorem

Expressions (1.15h) can also be written as a recurrence relationship with an index k running in the opposite direction to j

$$g_n(t) = h_n(t)$$

$$g_k(t) = h_k(t) * g_{k+1}(t), \quad k = n - 1, \ldots, 1 \tag{1.15i}$$

Given the impulse response, $h_k(t)$, of each sub-system in S_s, we can find the impulse response, $g_k(t)$, of the corresponding sub-system in S_p, by successive convolution in (1.15i) taken in reverse order using the index k. However, the inverse problem: "given S_p, find S_s", is considerably more difficult. Given the functions $g_k(t)$, we must solve $(n-1)$ problems of *de-convolution* to find all the $h_k(t)$ in expression (1.15i). Consequently, the *canonical form* for a linear time-invariant system, which can be decomposed into sub-systems, is the cascade form, S_s, consisting of sub-systems in series. For a particular case see Diskin and Pegram (1987).

De-convolution

Canonical form

1.4 DISCRETE FORMS OF CONVOLUTION EQUATION

In practice, input and output data are more frequently available in the form of either *instantaneous or cumulative inputs and outputs* sampled at a standard interval than in continuous form. In hydrology, rainfall data is most usually available as volumes over a standard interval and streamflow available from a continuous record. Under such circumstances it is possible to use the *quantized input data* directly in equation (1.3) by assuming the input to be uniform within the standard interval and to accept the error involved by this assumption.

Alternatively one can reformulate the basic equation to take account of the nature of the data to be used. For the case of a *quantized input* where the average input intensity during the interval from time $t = \sigma D$ to time $t = (\sigma + 1)D$ is denoted by $x(\sigma D)$, equation (1.10a) can be replaced by

$$y(t) = D \sum_{\sigma D = -\infty}^{\infty} x(\sigma D) \, h_D(t - \sigma D) \quad \sigma = \ldots, -1, 0, +1, \ldots \tag{1.16}$$

Pulse response

where $h_D(t)$ is the *pulse response* (i.e. the response of the system to a pulse of uniform input of length D) rather than the impulse response $h(t)$. Instead of using the *intensity* or *rate of inflow* $x(\sigma D)$ we can combine the value D, which appears in front of the integration sign in equation (1.16), with the rate of inflow to form the *volume of inflow over the interval* $X(\sigma D)$, in which case the equation will read

$$y(t) = \sum_{\sigma D=-\infty}^{\infty} X(\sigma D)\, h_D(t - \sigma D) \quad \sigma = \ldots, -1, 0, +1, \ldots \tag{1.17}$$

In the above equations, both the pulse response and the output are continuous functions of time. If the *output is sampled* at the standard interval D, then we have the completely discrete relationship

$$y(sD) = \sum_{\sigma D=-\infty}^{\infty} X(\sigma D)\, h_D(sD - \sigma D) \quad s, \sigma = \ldots, -1, 0, +1, \ldots$$

$$\tag{1.18}$$

where both σ and s are now discrete variables. If the *system* is *causal* and if the *input* is an *isolated* one, the infinite limits in equation (1.18) can be replaced by finite limits, as in the case of equation (1.13) for continuous time, thus becoming

$$y(sD) = \sum_{\sigma=0}^{s} X(\sigma D)\, h_D(sD - \sigma D) \quad s, \sigma = 0, 1, 2, \ldots \tag{1.19a}$$

which can be written without ambiguity as

$$y(s) = \sum_{\sigma=0}^{s} X(\sigma)\, h_D(s - \sigma) \quad s, \sigma = 0, 1, 2, \ldots \tag{1.19b}$$

in which the standard sampling interval D has been taken as the unit of time.

Because equation (1.19) is completely discrete in form, it represents a set of *simultaneous linear algebraic equations*. This set of equations can be written in matrix form as

$$\underline{y} = \underline{X}\,\underline{h} \tag{1.20}$$

where $\underline{y}$ is the column vector $(y_0, y_1, \ldots, y_{p-1}, y_p)$, formed from the known output ordinates sampled at interval D, $\underline{h}$ is the column vector of unknown ordinates of the pulse response $(h_0, h_1, \ldots, h_{n-1}, h_n)$, for the sampling interval D, and $\underline{X}$ is the matrix formed from the vector of input volumes $(X_0, X_1, \ldots, X_{m-1}, X_m)$ as follows

$$
\underline{X} = \begin{bmatrix}
X_0 & 0 & \cdot & \cdot & \cdot & 0 & 0 \\
X_1 & X_0 & \cdot & \cdot & \cdot & 0 & 0 \\
\cdot & \cdot & \cdot & \cdot & \cdot & \cdot & \cdot \\
\cdot & \cdot & \cdot & \cdot & \cdot & \cdot & \cdot \\
X_m & X_{m-1} & \cdot & \cdot & \cdot & \cdot & \cdot \\
0 & X_m & \cdot & \cdot & \cdot & X_0 & 0 \\
0 & 0 & X_m & \cdot & \cdot & X_1 & X_0 \\
\cdot & \cdot & 0 & \cdot & \cdot & X_2 & X_1 \\
\cdot & \cdot & \cdot & \cdot & \cdot & \cdot & \cdot \\
\cdot & \cdot & \cdot & \cdot & \cdot & X_m & X_{m-1} \\
0 & 0 & \cdot & \cdot & \cdot & 0 & X_m
\end{bmatrix}
\qquad (1.21)
$$

Problem of system
identification

The *problem of system identification* for a lumped, linear, time-invariant system is the problem of solving the above set of over-determined simultaneous linear equations for the unknown elements of the vector $\underline{h}$.

In applied hydrology, the solution of this identification problem is complicated by the fact that the input and output data are subject to *measurement error* and there may be further errors due to the approximations involved in the assumptions of linearity and time-invariance. The presence of such errors in the inversion problem becomes of serious significance when the mathematical system being inverted is an ill-conditioned one (Dooge and Bruen, 1989). Unfortunately, in the case of the heavily damped systems encountered in hydrology the output is a much smoother function than the input and consequently the mathematical inversion of this smoothing process is inherently unstable. Consequently, any proposed method for the solution of the set of equations, must be proved to be accurate not only for perfect data, but also robust when the data are subject to error.

1.5 SUGGESTIONS FOR FURTHER READING

As indicated above, the systems approach has been proposed in a wide variety of problem areas in recent years. A student is liable to be hindered rather than helped by the profusion of books and articles on the subject. The following notes are intending to help the student who wishes to read further on the subject. Details of the publications referred to are given at the end of this publication.

The use of a basic systems approach to the mathematical modelling of hydrologic processes and systems has been discussed by a number of authors particularly between 1960 and 1990. The general role of deterministic methods has been dealt with in papers by Amorocho and Hart (1964), Dooge (1968), Vemuri and Vemuri (1970), and in monographs by Becker and Glos (1969), Kuchment (1972), Dooge (1973) and Singh (1988). The general role of stochastic methods in systems hydrology has been dealt with in papers by Cavadias (1966), Kisiel (1969) and Clarke (1973) and in books by Svanidze (1964), Kartvelishvili (1967), Yevjevich (1972) and Bras and Rodriguez-Iturbe (1985). A study of the application of the systems approach to the design and operation of water resource systems could with advantage be started by reading such works as Maass and others (1962), Meta Systems (1975) and De Neufville (1990).

For those who wish to study the nature of the systems approach more deeply, it is useful to read of its application to disciplines other than hydrology. Many of the basic ideas of system theory as applied to mechanical and electrical systems are discussed clearly in such works as MacFarlane (1964), Shearer, Murphy and Richardson (1967), Martens and Allen (1969). Doebelin (1966) deals with the more special subject of instrumentation systems. Other special subjects on which books are available are chemical engineering systems (Franks, 1967) and estuarine and marine systems (Nihoul, 1975). Good introductions to the adaptive nature of economic, social and biological systems are contained in works by Tustin (1953), Bellman (1961), and Forrester (1968). A series of readings on "Systems thinking" covering biological and social systems was edited by Emery (1969).

General systems theory

The study of systems without reference to the nature of the system components is stressed in what is known as *general systems theory*. A good introduction to this subject is the book "An introduction to general systems thinking" by Weinberg (1975). More mathematical, but not as difficult to read as other books with the same objective, is "An approach to general systems theory" by Klir (1969) or Klir (1992). A collected series of papers by Bertalanffy, a biologist and a pioneer of general systems theory, have been published and make interesting reading (Bertalanffy, 1968).

CHAPTER 2

Nature of Hydrological Systems

2.1 THE HYDROLOGICAL CYCLE AS A SYSTEM

The hydrological cycle is usually depicted in a form similar to that shown in Figure 2.1, which is taken from a classical reference work by Ackerman et al. (1955). An alternative representation, which is more suitable for the present discussion, is shown in Figure 2.2. In the latter figure the rectangles denote various forms of water storage: in the atmosphere, on the surface of the ground, in the unsaturated soil moisture zone, in the groundwater reservoir below the water table, or in the channel network draining the catchment. The arrows in the diagram denote the various hydrological processes responsible for the transfer of water from one form of storage to another.

Water storage and water movement

A study of Figure 2.2 reveals the relationship between the various forms of water storage and water movement. Thus the precipitable water (W) in the atmosphere may be transformed by *precipitation* (P) to water stored on the surface of the ground. In the reverse direction, water may be transferred from the surface of the bare ground or from vegetation to the atmosphere by the processes of *evaporation* and *transpiration* (E, T).

Some of the water on the surface of the ground will infiltrate the soil through its surface (F) while some of it may find its way as *overland flow* (Q_O) into the channel network. During and following precipitation, soil moisture in the unsaturated sub-surface zone is replenished by *infiltration* (F) through the surface. If the *field moisture deficiency* of the soil, due to evaporation, transpiration and drainage since the previous precipitation event, is substantially satisfied by the current event, there will be a *recharge* (R) to groundwater, and also a certain amount of *interflow* (Q_i) or *lateral flow* through the soil, which is intercepted by the channel network.

The groundwater storage is depleted by groundwater outflow (Q_g), which enters the channel network and supplies the streamflow during dry periods. During prolonged drought, soil moisture may be replenished through *capillary rise* (C) from groundwater. Overland flow (Q_O), interflow (Q_i) and groundwater flow (Q_g) are all combined and modified in the channel network to form *the runoff* (RO) from the catchment. These various hydrological processes are discussed in detail in textbooks and monographs on physical hydrology (e.g. Eagleson, 1969; Bras, 1990) and are dealt with here in subsequent chapters.

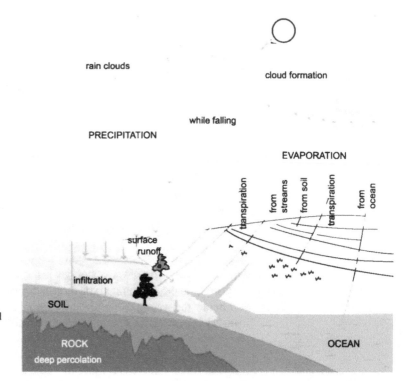

Figure 2.1. The hydrological cycle. A descriptive representation. (After Ackermann, Colman and Ogrosky, 1955.)

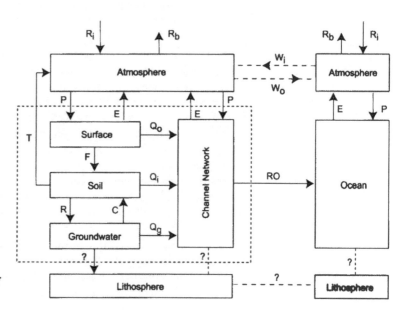

Figure 2.2. Block diagram of the hydrological cycle.

It is not easy in practice to distinguish from a record of runoff, the separate components of overland flow, interflow and groundwater flow discussed above. In the case of experimental plots, or of short-term event investigations, techniques using chemical or radioactive tracers can be used. However, these methods are not suitable for long-term routine monitoring of large catchments. The most that can be done is to distinguish *Rapid response* between the relatively rapid response of the catchment to a precipitation event and a second slower response. The quick response is often identified with surface runoff in the form of overland flow and interflow in the upper layers of the soil, and the slower response with the passage of the water through both the unsaturated and saturated soil moisture zones. Accordingly, most models of catchment behaviour used in applied hydrology are elaborations of the simplified catchment model shown in Figure 2.3. In this figure, the total response of the catchment to precipitation is due to three sub-systems: one representing *direct storm response*, the second representing groundwater storage and outflow, and the third representing the unsaturated soil moisture zone.

It would be desirable to model the overall catchment response shown in Figure 2.3 by a linear model. However, it is generally considered that recharge to groundwater (R) only takes place to an appreciable extent after the field moisture deficiency has been satisfied. This introduces into the *Threshold effect* catchment response a *threshold effect*. Since such an effect is essentially non-linear, it is not possible to model the total catchment response with a linear model without incorporating such a feature. Accordingly, it is not surprising that the first attempts at rainfall-runoff modelling concentrated on the direct storm response, where there is no such initial hindrance to the use of a linear model.

As will be shown below, the unit hydrograph approach introduced by Sherman (1932a) is based essentially on the assumption that the catchment converts *effective rainfall* to *storm runoff* in a linear time-invariant fashion. The *unit hydrograph* approach was used widely in applied flood hydrology for at least 25 years before its relationship to linear systems theory was realised and taken advantage of.

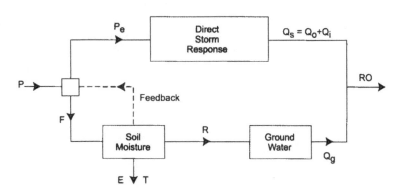

Figure 2.3. Simplified catchment model.

	Black box	Conceptual models	Continuum mechanics
Linear time-invariant	X	X	
Non-linear time-invariant	X	X	
Linear time-variant			
Non-linear time-variant			

Figure 2.4. Models of hydrological processes.

If the recharge to the groundwater sub-system is known, then there is no reason why this component should not be treated in a similar fashion. However, it was also relatively late when Kraijenhoff van de Leur (1958, 1966) did this. Because the techniques were developed firstly in regard to the component of direct storm response, there will be a tendency in this book to cite examples from this area. But it should be realised that all of these techniques are equally applicable to groundwater flow.

Mention was made in Chapter 1 of the three distinct approaches to the analysis of systems:

(1) black-box analysis,
(2) conceptual models,
(3) equations of continuum mechanics.

Any one of these approaches can be applied to each of the components of the total catchment response shown in Figure 2.3. In each approach one may assume the sub-system to be linear or non-linear, and to be time-invariant or time-variant. Without further classification, this divides possible models of independent components of the catchment response, or individual hydrological processes, into twelve classes as indicated in Figure 2.4. In the present book it is not possible to deal with all or even with a majority of these classes. Accordingly attention will be concentrated on *linear time-invariant black-box analysis*, *linear conceptual models* and *non-linear conceptual models*, i.e. on the upper left-hand corner of the matrix shown in Figure 2.4.

Unit hydrograph

The remainder of the present chapter will be devoted to a review of classical *unit hydrograph* theory and of the early classical methods for the identification and simulation of hydrological systems. The following chapter is devoted to a discussion of some of the mathematical tools required for understanding the deterministic methods used in systems hydrology.

2.2 UNIT HYDROGRAPH METHODS

The unit hydrograph concept and its development were one of the highlights of the classical period of scientific hydrology. Though unit

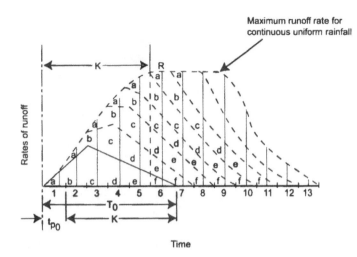

Figure 2.5. Superposition of unit hydrographs (after Sherman, 1932a).

hydrograph methods are dealt with elsewhere in this book, the approach is briefly reviewed here so that the black-box approach to hydrological systems can be placed in its proper historical context.

Figure 2.5 is a reproduction of Figure 1 of the basic paper by Sherman (1932a), which introduced the concept of the *unit hydrograph*. In this figure a triangular form of hydrograph is assumed to represent the *direct storm runoff due to effective rain falling continuously and uniformly for the unit interval*. The figure shows how superposition may be used to build up the runoff of periods of uniform rainfall equal to twice the unit period, three times the unit period, etc. It will be noted that, if the duration of this continuous effective precipitation is greater than the base of the unit hydrograph, the runoff becomes constant.

As mentioned above, the unit hydrograph methods were widely used in applied hydrology for about 25 years without recognition of the essential assumption involved, namely that of a linear time-invariant system.

The classical unit hydrograph approach is described in a number of textbooks published at the end of the 1940s. The book by Johnstone and Cross (1949) contains a particularly good discussion from this classi-
Basic propositions cal period. In it they state the basic propositions of the unit hydrograph approach as follows:

"We are now in a position to state the three basic propositions of unit graph theory, all of which refer solely to the surface-runoff hydrograph:

I For a given drainage basin, the duration of surface runoff is essentially constant for all uniform-intensity storms of the same length, regardless of differences in the total volume of the surface runoff.

II For a given drainage basin, if two uniform-intensity storms of the same length produce different total volumes of surface runoff, then the rate of surface runoff at corresponding times t, after the beginning of the two storms are in the same proportion to each other, as the total volumes of the surface runoff.

III Time distribution of surface runoff from a given period is independent of concurrent runoff from antecedent storm periods."

Johnstone and Cross make the following comment:

"All these propositions are empirical. It is not possible to prove them mathematically. In fact, it is a rather simple matter to demonstrate by rational hydraulic analysis that not a single one of them is mathematically accurate. Fortunately, nature is not aware of this."

In the twenty years after Sherman's basic paper of 1932 the unit hydrograph approach developed into a flexible and useful tool in applied hydrology and only later was a full theory of the unit hydrograph developed.

The original unit hydrograph developed by Sherman was a continuous runoff hydrograph due to uniform rainfall in a unit period. A later advance in classical unit hydrograph theory was the discovery by W. Langbein that the S-hydrograph or S-curve could be used to convert a unit hydrograph from one duration to another. Before this, it was necessary to find a storm of the appropriate duration in the period of record in order to derive the required unit hydrograph from the data.

S-hydrograph

The *S-hydrograph* or *S-curve* is defined as the *hydrograph of surface runoff produced by continuous effective rainfall at a constant rate*. Figure 2.5 (taken from Sherman's original paper) shows the manner in which the S-hydrograph can be built up from the unit hydrograph for the particular shape of a triangular unit hydrograph. Using a common time axis, the corresponding blocks of rain are frequently plotted upside down above the hydrograph. These are not shown in Figure 2.5.

Once the S-hydrograph has been adequately defined on the basis of a given unit hydrograph of any specified unit period, a unit hydrograph of a new unit period (D) can be derived from it. We simply displace the S-hydrograph by the amount D, subtract the ordinates of the two S-hydrographs, and normalise the volume by dividing by D. This process can be represented by the equation

$$h_D(t) = \frac{S(t) - S(t - D)}{D} \tag{2.1}$$

As D becomes smaller and smaller the process represented by equation (2.1) approaches closer and closer to the definition of differentiation and in the limit we have

$$h_0(t) = \frac{\mathrm{d}}{\mathrm{d}t}(S(t)) \tag{2.2}$$

Instantaneous unit hydrograph (IUH)

The hydrograph defined by equation (2.2) is known as the *instantaneous unit hydrograph* (IUH).

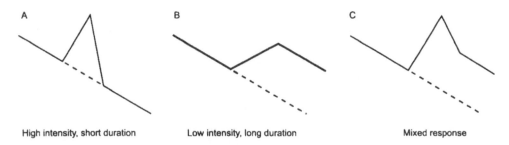

High intensity, short duration   Low intensity, long duration   Mixed response

Figure 2.6. Hydrograph response (after Horton, 1935).

Storm event

The first step in analysing an actual hydrograph is to separate the baseflow from the *direct storm runoff* and to reduce the total precipitation to *effective precipitation*. Hydrological literature abounds in methods for making this separation. The effect of contrasting types of storm event on the runoff hydrograph is shown schematically in Figure 2.6 due to Horton (1935).

Figure 2.6A shows the effect of a storm event with very high intensity of rainfall and very short duration. Because of the very high intensity of rainfall, surface runoff would occur due to the exceedance of the limiting or maximum rate of infiltration. But due to the very short duration, and consequently the very small volume of infiltration, it is unlikely that the field moisture deficit would be satisfied. Hence, there would be no recharge to groundwater. Under these conditions, the baseflow recession before and after the storm event would follow the same general curve, and the response of the hydrograph would consist of a sharp rise and sharp recession back to the same master curve of baseflow recession.

If on the other hand the storm event consisted of very prolonged rainfall of very small intensity, we would get the condition represented schematically in Figure 2.6B. In this case the intensity does not exceed the maximum infiltration rate and thus there is no surface runoff. However, the rainfall is sufficiently prolonged to make up the field moisture deficiency and to give a recharge to groundwater storage. The effect of this recharge is to increase the amount of groundwater outflow, or baseflow, and the recession curve will be shifted as shown in stylised fashion in Figure 2.6B. In the latter case, the recession curve after the storm event has the same shape as before, but is shifted in time.

More usually, the storms which are of consequence in hydrological analysis, are intermediate between the two extreme forms discussed above. Both effects are combined, so that we get, on the one hand, a distinct peakflow and a measurable amount of surface runoff, and on the other hand, a recharge of groundwater, giving a shift in the master recession curve. This mixed condition is shown schematically in Figure 2.6C. One of the first steps necessary in unit hydrograph analysis is to separate out these two effects.

In practice, the applied hydrologist usually separates the baseflow in some arbitrary fashion, and then adjusts the total precipitation so that the volume of effective precipitation is equal to the volume of direct storm runoff. No attempt is made to link infiltrating precipitation with groundwater recharge and hence with ground-water outflow.

The reduction of total precipitation to effective precipitation is also frequently made in an arbitrary manner. Figure 2.7 shows three approaches to the adjustment of the precipitation pattern.

Infiltration capacity curve

In the first method, an *infiltration capacity curve*, which decreases as the soil moisture is satisfied, is estimated, and an allowance is also made for the retention of precipitation on vegetation and on the surface of the ground without the occurrence of runoff.

W-index method

In the second method (which is called the *W-index method*), the difficulty of estimating the variable infiltration capacity is avoided, by taking it to be a constant value which will make the volume or effective precipitation equal to the volume of direct storm runoff after allowance for retention.

φ-index method

In the third method (known as the *φ-index method*) no allowance is made for retention and we simply seek the value of infiltration capacity such that the volume of precipitation in excess of it, is equal to the volume of surface runoff. It is clear that the use of either of the latter two methods, or the incorrect use of the first method, would result in the derivation of an erroneous unit hydrograph, even if the total precipitation and total runoff were measured without error and the catchment truly acted in a linear and time-invariant fashion.

All of the concepts and methods of the classical unit hydrograph approach can be neatly formulated in systems nomenclature. The only necessary assumptions of the unit hydrograph approach are those of linearity and time-invariance (Dooge, 1959). In the classical statement of unit hydrograph principles by Johnstone and Cross (1949) as quoted above in Section 2.3, the basic propositions involve both these two concepts.

Linearity and time-invariance

Proposition I invokes time-invariance when it compares storms occurring at different times and invokes linearity when it assumes the base length of the unit hydrograph to be constant. Proposition II again invokes time-invariance by comparing two storms and invokes linearity in assuming proportionality of the runoff ordinates. Proposition III is an alternative statement of the principle of superposition i.e. of linearity applied to two successive storms.

The above considerations show that linearity and time-invariance are necessary for the basic propositions given by Johnstone and Cross. A comparison indicates that the two assumptions of linearity and time-invariance are also sufficient for the three basic propositions.

The unit hydrograph for a unit duration ($D = 1$) is clearly the *pulse response* for that duration and the instantaneous unit hydrograph is the *impulse response* of the catchment. The S-hydrograph corresponds to the *step-function response* of system theory.

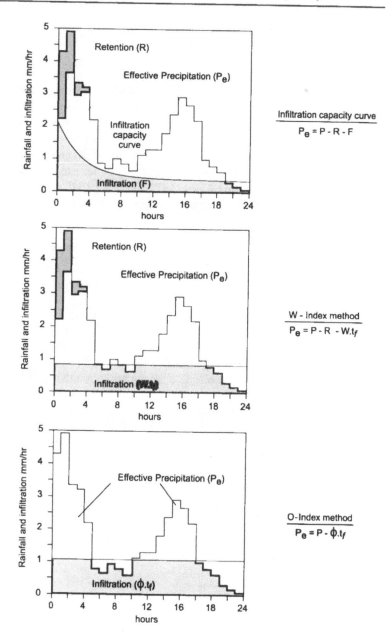

Figure 2.7. Effective precipitation.

Separation of baseflow

Effective precipitation

The problems of the separation of baseflow and the determination of effective precipitation is a reflection of the fact that the direct storm response is merely one component of the total catchment response as shown above in Figure 2.3. The systems approach, however, emphasises the undesirability of a partial approach. The failure to link the portion of the precipitation, which infiltrates, with the groundwater recharge, may give results, which are not only inaccurate, but also grossly inconsistent and erroneous.

Once the problem has been formulated in systems terms, it is clear that the problem of deriving a unit hydrograph is one of solving an integral equation (1.15) in the continuous case and solving a set of simultaneous linear equations (1.20) in the discrete case. There have been a number of attempts to estimate simultaneously the unit hydrograph itself and one or other of the "slow response" components. In the case of the effective precipitation function, see for example, the CLS method used by Todini and Wallis (1977), and the iterative solution of problems of identification and detection, presented by Sempere Torres, Rodriguez and Obled (1992).

2.3 IDENTIFICATION OF HYDROLOGICAL SYSTEMS

In the classical unit hydrograph approach it was found necessary to devise methods for the derivation of a unit hydrograph by the analysis of complex storms in which the rate of effective precipitation was not uniform throughout the storm. In the early studies the procedure used was essentially one of *trial and error*. In this method, which is shown schematically in Figure 2.8, a shape of unit hydrograph was assumed and applied to the effective precipitation in order to produce an estimate of the runoff. This was then compared with the recorded runoff. If the fit was felt to be adequate, the shape of the unit hydrograph was adopted. But if not, it was modified until satisfactory agreement was obtained.

Iterative method   In 1939, Collins suggested an *iterative method*. A trial unit hydrograph, generated by applying the assumed unit hydrograph to all periods of rainfall except the maximum, was subtracted from the total measured hydrograph to give a net runoff hydrograph, which could be taken as the runoff due to the maximum rainfall in a unit period. This net hydrograph,

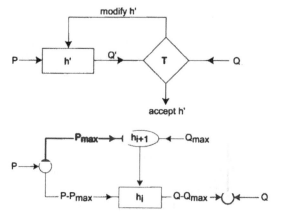

Figure 2.8. Classical methods of unit hydrograph derivation.

when divided by the volume of the maximum unit period rainfall, gave a second approximation to the shape of the finite period unit hydrograph. The process was repeated until there was a satisfactory correspondence between the shapes of unit hydrograph obtained on the two successive iterations.

In the 1940s the derivation of the unit hydrograph from complex storms was frequently based on the solution of the set of simultaneous equations giving the relationship between sampled ordinates of the output and the ordinates of the finite period of the unit hydrograph and the rainfall volumes in each unit period (Linsley, Kohler and Paulhus, 1949, pp. 444–449). These were of course the set of equations represented in matrix form by equation (1.20) above.

When written out explicitly they have the form

$$y_0 = x_0 h_0 \tag{2.3a}$$

$$y_1 = x_1 h_0 + x_0 h_1 \tag{2.3b}$$

$$y_2 = x_2 h_0 + x_1 h_1 + x_0 h_2 \tag{2.3c}$$

and so on. Note the change in notation compared with (1.19b). Subscripts are used instead of arguments and lower case x denotes successive input volumes in this chapter.

The general form of the equation is

$$y_i = x_i h_0 + x_{i-1} h_1 + \cdots + x_0 h_i \tag{2.3d}$$

and the number of terms on the right hand side of the equation increase until it is equal to either the number of elements in the input vector $(m+1)$ or the number of elements in the pulse response vector $(n+1)$, whichever is the lesser. For the case of $m < n$, this equation is given by

$$y_m = x_m h_0 + x_{m-1} h_1 + \cdots + x_0 h_m \tag{2.3e}$$

and the next equation will be the same length, since $x_{m+1} = 0$ and does not contribute to it, thus giving

$$y_{m+1} = x_m h_1 + \cdots + x_0 h_{m+1} \tag{2.3e'}$$

The equations will remain the same length until $i = n$, after which $h_i = 0$ and each successive equation will contain one less term

$$y_n = x_m h_{n-m} + \cdots + x_0 h_n \tag{2.3f}$$

$$y_{n+1} = x_m h_{n+1-m} + \cdots + x_1 h_n \tag{2.3f'}$$

The equations continue to decrease until ultimately we get

$$y_p = x_m h_{p-m} = x_m h_n \tag{2.3g}$$

which is the last equation of the set.

Forward substitution

It would appear that the solution of the above equations is trivial since we can proceed by *forward substitution*. In forward substitution, h_0 is the only unknown in equation (2.3a) and hence can be determined from the values of y_0 and x_0. Equation (2.3b) can then be solved for h_1 using the known values of x_0, x_1 and y_1 and the value of h_0 derived from equation (2.3a). In practice, the existence of errors in the values of the effective precipitation $x(sD)$ or the direct runoff $y(sD)$ will produce errors in the ordinates of the unit hydrograph. Applied hydrologists knew that for certain patterns of effective precipitation, these errors could build up rapidly and produce unreasonable results.

Method of least squares

One method proposed to overcome this disadvantage was the solution of the equations represented by equation (2.3) by the *method of least squares*. This method of unit hydrograph derivation will be discussed in greater detail in Chapter 4. Barnes (1959) removed any oscillations occurring in the unit hydrograph by deriving it in the reverse order i.e. by backward substitution. This is in line with general experience in numerical methods where a calculation, which is unstable in one direction, is usually stable if taken in the reverse direction. Barnes further suggested that the estimated effective precipitation should be adjusted until the unit hydrograph obtained in the forward and reverse directions was substantially the same. Later attempts at developing objective and dependable methods of unit hydrograph derivation, involving the use of *transform methods* or of *optimisation methods* both unconstrained and constrained, will be discussed later in Chapter 4.

Transform methods

Optimisation methods

Although the derivation of the unit hydrograph from the outflow hydrograph due to a complex storm (which is the problem of identification) is a difficult one to solve, the *prediction* of the flow hydrograph due to a complex storm, when the unit hydrograph is known, is relatively easy. All that is required is the application of each of the volumes of effective precipitation in a unit period to the known finite period hydrograph. To obtain the outflow hydrograph each volume of effective precipitation must be carefully located in time and the results summed to give the total outflow. In terms of the set of simultaneous equations given by equation (2.3a) to (2.3g), the problem is one of determining the left hand side of the equations knowing all the values of $\underline{x}$ and all the values of $\underline{h}$, which appear on the right hand sides of the equations.

2.4 SIMULATION OF HYDROLOGICAL SYSTEMS

Even if we could solve the problem of identification completely, this would only enable us to predict the future output from the particular system for which the identification was made. Complete and accurate identification would not help us in any way to predict the output from a similar system, for which records of input and output were not available, or to study the effects

Simulation

of variations in the parameters due to natural causes or human activity. Furthermore, the black-box analysis of non-linear systems is extremely difficult and there is still no adequate theory to deal with the analysis of systems containing thresholds. Consequently, in order to predict the total catchment response, and, in some cases, the response of catchment components, it is necessary to turn to *simulation* as the basis of prediction. It is important to remember that in these cases, we are still interested in the overall performance of the system rather than the details. When we seek a model to simulate the system operation, we are still looking for a reliable predictor, rather than a fundamental explanation, or a detailed reproduction of the system.

Direct approach to simulation

There are of course many methods of simulation. On the one hand we have direct methods of simulation, which relate to a particular hydrological system. These include the use of physical models in hydrology and the use of special purpose analogs. The discussion of this *direct approach* to simulation is outside the scope of the present chapter. In contrast to direct simulation, are the formulation of a mathematical model and the solution of this model either by hand calculation, use of a desk calculator, or digital computer. In this book, we are concerned only with the mathematical modelling of a deterministic relationship between input and output variables.

Regression models

Regression models have been widely used in applied hydrology down through the years. The value of regression models is clearly their ability to predict, rather than the investigation of the details of hydrological processes. A typical example of a regression model is shown on Figure 2.9 (Benson, 1962). In this particular regression model, a relationship is sought between the annual peak discharge for a recurrence interval of T years (Q_T) as the dependent variable, and a number of catchment parameters as independent variables. In the case shown, the catchment parameters are:

the catchment area (A), the main channel slope (S), a measure of the surface storage area (S_t), the 24-hour rainfall for a recurrence period of T years (I), a measure of freezing conditions in mid winter (t), and an orographical factor (O).

In Figure 2.9 the particular regression model is shown in three forms:

(1) as an indeterminate relationship between the dependent variable and the independent variables;
(2) as an equivalent flow diagram, indicating that each of the independent variables can be considered as input factors, which are raised to a particular power and then multiplied together, to give the required output; and
(3) as a regression model, expressed as a relationship between the logarithm of the dependent variable, and the logarithms of the independent

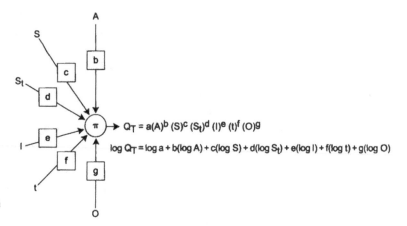

Figure 2.9. Typical regression
model.

$$Q_T = a(A)^b (S)^c (S_t)^d (I)^e (t)^f (O)^g$$

$$\log Q_T = \log a + b(\log A) + c(\log S) + d(\log S_t) + e(\log I) + f(\log t) + g(\log O)$$

variables. In this way the relationship is transformed, so as to be linear in the unknown parameters, which can therefore be found by linear regression analysis.

Even when the unit hydrograph approach is used, it is necessary to establish a regression model relating the unit hydrograph parameters to catchment characteristics. *Synthetic unit hydrographs* are required for catchments with no flow records, and are formulated on the basis of catchments in the same region for which records are available. Regression models are also useful for investigating the variation of unit hydrograph characteristics with effects of human activity, such as changes in agricultural practice or urbanisation.

Synthetic unit hydrographs

Multiple regression analysis makes the assumption that all the errors are concentrated in the dependent variable and also that the so-called independent variables are not correlated with one another. Violation of the latter assumption does not prevent the derivation of a regression relationship, which can be used as a prediction tool, but it renders meaningless the tests of significance used in a regression analysis. In hydrology, due to the operation of geomorphologic factors, the catchment parameters are often very highly correlated among themselves. Methods of *multivariate analysis* seek to avoid the above problems by treating all the variables alike and performing component analysis to determine if there are truly independent groupings of variables.

Multivariate analysis

The graphical method of *coaxial correlation* was widely used in the past in applied hydrology (Linsley, Kohler and Paulhus, 1949, pp. 419–424). It is essentially a graphical method of non-linear regression and is suitable for the solution of *ad hoc* problems. One of the disadvantages of coaxial correlation is the subjectivity involved. Becker (1966) has indicated how the subjectivity can be substantially reduced by relating the shape of the curves in the coaxial correlation diagram to a given

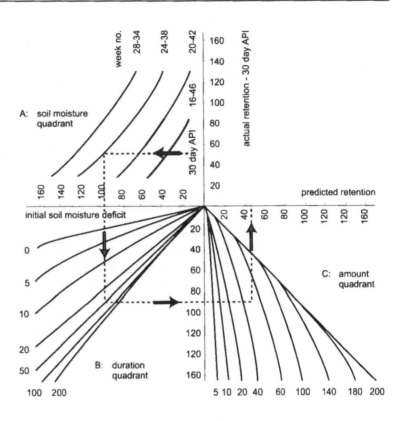

Figure 2.10. Coaxial correlation diagram (Becker, 1966).

conceptual model of catchment behaviour. Figure 2.10 is an example of a coaxial correlation diagram based on Becker's work which relates potential catchment recharge to initial moisture content, time of year, duration and amount of rainfall.

Total catchment response

In order to build up a model of *total catchment response*, it is necessary to model separately the components of this response. As indicated earlier, this may be done either by black-box analysis, by the use of simple conceptual models, or by solution of the equations of continuum mechanics. In this book, only black-box analysis and simple conceptual models will be considered. It was noted above that the unit hydrograph method had been used widely in applied hydrology, long before the development of a theory of black-box analysis of hydrological system. Similarly, simple conceptual models were devised for use in applied hydrology long before system simulation was applied to hydrology in a systematic fashion.

Muskingum method of flood routing

A good example of this, is the well-known *Muskingum method of flood routing*. This method assumes that the storage in a river reach can be represented by

$$S(t) = K[xI(t) + (1 - x)Q(t)] \tag{2.4}$$

where $S(t)$ is the storage, $I(t)$ is the inflow at the upstream end of the reach, $Q(t)$ is the outflow at the downstream end of the reach, and x and K are parameters to be determined. For the particular case of $x = 0$ equation (2.4) reduces to the equation for a linear reservoir, which is widely used as a basis for building simple conceptual models. It will figure prominently in later chapters.

Linear reservoir

In classical hydrology, the Muskingum parameters were determined as follows. From the record of I and Q for an existing flood, the value of the storage $S(t)$ in the reach is obtained from the hydrological continuity equation

$$\frac{d}{dt}[S(t)] = I(t) - Q(t) \tag{2.5}$$

by the following empirical procedure.

Various values of x between 0 and 0.5 were selected and the weighted flow in the reach — i.e. the quantity within square brackets in equation (2.4) — calculated and plotted against the storage. The value of x selected, was the value which gave the least dispersion about a straight line in a graphical plot of storage against weighted flow. A line was drawn through the plotted points for this particular value of x. The slope of this line determines the value of the parameter K, and it has the dimension of time. The values of x and K thus found were used to determine the outflow from the reach for a new design inflow I.

It may be instructive to reinterpret the above procedures in terms of systems nomenclature. The Muskingum method is clearly a method of simulation based on a simple conceptual model, rather than the solution of the St. Venant equations for unsteady flow in an open channel. For the case of a natural reach of a river, the solution of the latter equation is difficult, even with the aid of a digital computer. It is interesting to note that, even if the equations can be satisfactorily solved on a computer, it is necessary to use an existing record of water level at two points, which may be converted to flow, in order to determine reliable values of the "friction" factor to be used.

St. Venant equations

The graphical procedure for the determination of x and K in the classical Muskingum method is essentially a method of identifying the values of the parameters of the Muskingum model for the special case under consideration. It constitutes therefore *parameter identification* for the chosen conceptual model. The use of the model and these values of the parameters to estimate the design outflow for a design inflow, represent the relatively simple problem of *prediction* using the derived river reach response. The form of equation (2.4) was based on heuristic assumptions about relative importance of *reservoir storage* and *wedge storage* in the reach. In the systems approach no use would be made of such physical assumptions; instead, equation (2.4) would be interpreted as an assumption that the storage in the reach is a linear function of the inflow and the outflow. The

Parameter identification

Reservoir storage

Wedge storage

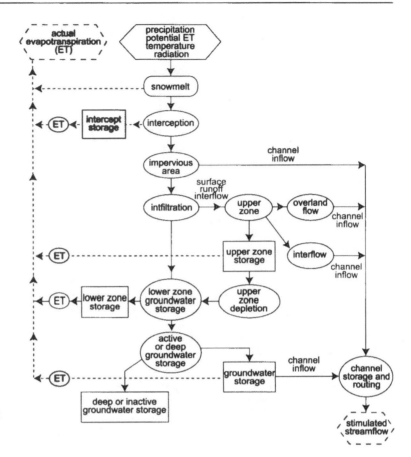

Figure 2.11. Stanford Model
Mark IV (Crawford and
Linsley, 1966).

behaviour of this linear model can then be characterised by its impulse
response function. The impulse response of the Muskingum model is a
linear combination of a negative delta function at the origin and a negative exponential function of time. Under certain conditions, the negative
delta-function can produce negative outflows from the reach. This is a
significant weakness in this popular model.

Figure 2.11 shows an outline flow diagram for the older Stanford Mark
IV model for simulating an entire catchment (Crawford and Linsley, 1966).
This is not a complete description of the model and needs to be supplemented by a series of instructions indicating how water is to be transferred
from one element of the model to another.

An alternative way of representing such a conceptual model is shown
in Figure 2.12 which presents the model used by Dawdy and O'Donnell
(1965) to investigate the problem of developing objective and efficient
techniques for optimising the parameters of a model.

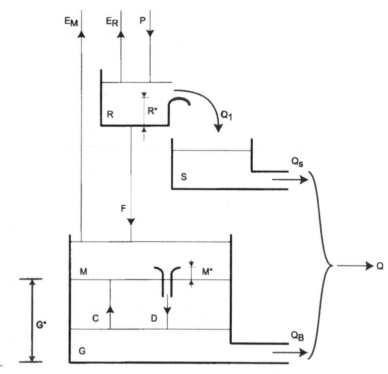

Figure 2.12. Schematic
diagram of the overall model
of the hydrological cycle.
(Dawdy and O'Donnell, 1965).

*Total catchment
response*

The discussion of the problem of the simulation of total catchment
response is outside the scope of this book. A good starting point however
is provided by the review of Franchini and Pacciani (1991). Many conceptual and physically-based lumped models of total catchment response are
described in the literature and all three components shown in Figure 2.3
and their interconnection with one another are treated (Singh et al., 2002).

TOPMODEL (Beven et al., 1984), SHE (Abbott et al., 1986), the
XIANJIANK and ARNO models (Todini, 1988 and 1996) and TOPKAPI
(Todini, 1995; Ciarapica and Todini, 1998; Todini and Ciarapica, 2001)
are examples of semi-distributed and distributed models of the total catchment. Some of these models require large data sets for the calibration of
their distributed parameter fields. The dependence of such fields on the
model grid-scale and the parameterisation of sub-grid-scale processes are
active areas of research.

CHAPTER 3

Some Systems Mathematics

3.1 MATRIX METHODS

Matrix

A matrix is an array or table of numbers. Thus we define the matrix $\underline{A}$ as

$$
\underline{A} = \begin{bmatrix}
a_{11} & a_{12} & \cdot & \cdot & \cdot & a_{1n} \\
a_{21} & a_{22} & \cdot & \cdot & \cdot & a_{2n} \\
\cdot & & \cdot & \cdot & & \cdot \\
\cdot & & & \cdot & & \cdot \\
a_{m1} & a_{m2} & \cdot & \cdot & \cdot & a_{mn}
\end{bmatrix} \tag{3.1}
$$

This matrix which has m rows and n columns is referred to as an $m \times n$ matrix. The figure $\underline{A}$ is used as a mathematical shorthand for the table of numbers on the right hand side of equation (3.1). Matrix algebra tells us what rules should be used to manipulate such arrays of numbers.

If a matrix $\underline{C}$ is composed of elements, each of which is given by adding the corresponding elements of matrix $\underline{A}$ and matrix $\underline{B}$, that is:

$$
c_{ij} = a_{ij} + b_{ij} \tag{3.2}
$$

the matrix $\underline{C}$ is said to be the sum of the two matrices $\underline{A}$ and $\underline{B}$ and we write:

$$
\underline{C} = \underline{A} + \underline{B} = \underline{B} + \underline{A} \tag{3.3}
$$

Matrix multiplication is defined as a result of the operation:

$$
\underline{C} = \underline{A} \cdot \underline{B} \tag{3.4}
$$

where the elements of $\underline{C}$ are defined as

$$
c_{rt} = \sum_{s} a_{rs} b_{st} \tag{3.5}
$$

It is essential for an understanding of matrix operations to see clearly the nature of the operation defined by equation (3.5). The element at the intersection of the r row and t column in the $\underline{C}$ matrix is obtained by multiplying term by term the r row of the $\underline{A}$ matrix by the t column of the $\underline{B}$ matrix and summing these products. This definition implies that matrix $\underline{A}$ has the same number of columns as matrix $\underline{B}$ has rows. It must be realised that in general:

$$\underline{A} \cdot \underline{B} \neq \underline{B} \cdot \underline{A} \tag{3.6}$$

i.e. that matrix multiplication is in general non-commutative.

A certain amount of nomenclature must be learnt in order to be able to use matrix algebra. When the numbers of rows and columns are equal the matrix is said to be *square* and if all the elements other than those in the principal diagonal (from top left to bottom right) are zero the matrix is called a *diagonal matrix*. A diagonal matrix in which all the principal diagonal elements are unity is called the *unit matrix* $\underline{1}$. The unit matrix $\underline{1}$ serves the same function as the number 1 in ordinary algebra and it can be verified that the multiplication of any matrix by the unit matrix gives the original matrix. An *upper triangular matrix* is one with non-zero elements on the principal diagonal and above, but only zero elements below the main diagonal. A *lower triangular matrix* has non-zero elements in the principal diagonal and below it, but only zero elements above the main diagonal.

Unit matrix

The matrix whose *ij*-th element a_{ij}, is a function of $(i - j)$, rather than of i and j separately, is called a *Toeplitz matrix*. A Toeplitz matrix of order 4, for example, is

$$\begin{bmatrix} a_0 & a_{-1} & a_{-2} & a_{-3} \\ a_1 & a_0 & a_{-1} & a_{-2} \\ a_2 & a_1 & a_0 & a_{-1} \\ a_3 & a_2 & a_1 & a_0 \end{bmatrix} \tag{3.7}$$

The *transpose* $\underline{A}^T$ of a matrix $\underline{A}$ is the matrix, which is obtained from this original matrix by replacing each row by the corresponding column and vice versa. If the transpose of the matrix is equal to the original matrix then the matrix is said to be *symmetrical*. The individual rows and columns of a matrix may be considered as row vectors and column vectors. The transpose of a row vector will be a column vector and vice versa. The *inverse* of a matrix $\underline{A}$ is a matrix $\underline{A}^{-1}$, which when multiplied by the original matrix $\underline{A}$ gives the unit matrix $\underline{1}$ that is:

$$\underline{A} \cdot \underline{A}^{-1} = \underline{I} = \underline{A}^{-1} \cdot \underline{A} \tag{3.8}$$

The transpose (or the *inverse*) of a *matrix product* is equal to the matrix product of the transposes (or *inverses*) of the basic matrices but taken in reverse order. A matrix will only possess an inverse if it is square and is *non-singular* i.e. if its determinant is not equal to zero. A matrix is said to be *orthogonal* if its matrix is equal to its transpose, that is:

$$A^T = A^{-1} \tag{3.9}$$

Thus an orthogonal matrix has the great advantage that the potentially unstable process of inversion is replaced by the stable process of transposition.

A set of simultaneous linear algebraic equations are represented in matrix form by

$$\underline{A}\,\underline{x} = \underline{b} \tag{3.10}$$

where $\underline{A}$ is the matrix of coefficients, $\underline{x}$ is the vector of unknowns and $\underline{b}$ is the vector of the right hand sides of the *simultaneous equations*. If the number of equations is equal to the number of unknowns, the matrix of coefficients $\underline{A}$ will be square matrix; and, if it is non-singular it will also possess an inverse. The formal solution to the set of equations can therefore be obtained by multiplying each side of equation (3.9) on the left-hand side by the inverse $\underline{A}^{-1}$ thus obtaining:

$$\underline{A}^{-1}\underline{A}\,\underline{x} = \underline{x} = \underline{A}^{-1}\underline{b} \tag{3.11}$$

Ill-conditioned

From the point of view of actual computation, a matrix may be non-singular, but may still give rise to difficulty, because the equations are ill-conditioned resulting in a matrix which is almost singular, so that numerical results may become unreliable. Computer packages are available for the inversion of matrices and for the solution of simultaneous equations by both direct and iterative methods.

For further information on matrices and matrix solution of equations see Korn and Korn (1961), Bickley and Thompson (1964), Frazer, Duncan and Collar (1965), Raven (1966), and Rektorys (1969).

3.2 OPTIMISATION

Optimisation techniques can be applied both in the black-box analysis of systems and in the parameter identification of conceptual models. It will be first discussed in relation to the black-box analysis of the system identification problem. It has already been pointed out that when the input and output data are available in discrete form, the problem of system identification reduces to the problem of solving the sets of simultaneous linear algebraic equations represented by equation (1.20). In this set of equations there will be more equations than unknowns, since there are $(p+1)$ ordinates of sampled runoff, $(m+1)$ ordinates representing quantised rainfall and $(n+1)$ ordinates of the unknown pulse response. These subscripts are, by definition, connected by the equation

$$p = m + n \tag{3.12}$$

Overdetermined equations

which shows the degree to which the equations are *overdetermined*. There are m *redundant* equations. Consequently it is not possible to invert the matrix $\underline{X}$ in equation (1.20) in order to obtain a direct solution. Selecting different sub-sets of $(n + 1)$ equations may lead to *contradictory results*.

One approach to these difficulties is to seek the vector of pulse response ordinates that will *minimise the sum of the squares of the residuals*: the differences between the output predicted using this pulse response and the recorded output.

Thus if we write

$$\underline{r} = \underline{y} - \underline{X}\,\underline{h} \tag{3.13}$$

the sum of the squares of the residuals thus defined will be given by

$$\sum r_i^2 = \underline{r}^T \underline{r} \tag{3.14}$$

Using equation (3.12) and the rule for the transpose of a product, we can write

$$\sum r_i^2 = (y^T - h^T X^T)(y - Xh) \tag{3.15}$$

Expanding the right hand side of equation (3.14) gives

$$\sum r_i^2 = y^T y - y^T Xh - h^T X^T y + h^T X^T Xh \tag{3.16}$$

It should be noted that the sum of the squares of the residuals will be a scalar i.e. a one by one matrix and hence every element of equation (3.16) must be a scalar. Since the transpose of the scalar gives itself, the second and third terms on the right hand side of equation (3.16) must be identical.

The optimum least squares vector $\underline{h}$ will be that vector which minimises the sum of the squares of the residuals as given by equation (3.13). Advantage can be taken of matrix methods in order to differentiate the equation with respect to the vector $\underline{h}$ rather than with respect to each individual element (h_I) of it. Naturally the result is the same, the only difference being that the use of vector differentiation is more compact. Accordingly, we differentiate equation (3.15) with respect to the vector $\underline{h}$

$$\frac{\partial}{\partial \underline{h}}\left(\sum r_i^2\right) = -2\underline{X}^T \underline{y} + 2\underline{X}^T \underline{X}\,\underline{h} \tag{3.17}$$

and set the result equal to zero thus obtaining as the equation for the optimum vector pulse response ordinates

Optimum response

$$(\underline{X}^T\underline{X})\underline{h}_{opt} = \underline{X}^T \underline{y} \tag{3.18}$$

Since the matrix $(\underline{X}^T\underline{X})$ is of necessity a square matrix, it can (provided it is not singular) be inverted as in equation (3.11) above, to give a solution for the optimum vector $\underline{h}$ which will minimise the residuals between the predicted and measured outputs. $(\underline{X}^T\underline{X})$ is also a Toeplitz matrix and very efficient techniques have been developed for solving (3.18) (Zohar, 1974).

The classical least squares optimisation outlined above, represents unconstrained optimisation. Its application to real systems may result in an optimum vector which has properties that are considered unrealistic for the type of system being analysed. Thus in the case of the identification of the hydrological system represented by equation (2.3), the *continuity of mass* requires that the volumes of effective rainfall and direct storm runoff should be equal, and hence that the area under the impulse response should be one, and that the sum of the ordinates of the pulse response should be equal to unity.

Similarly, the application of the least squares method to data subject to measurement error might result in a solution that is far less smooth than would be expected for the type of system being examined. It is possible to extend the least squares system and develop techniques for *constrained optimisation*. In these methods, we seek either the vector that minimises the residuals subject to the satisfaction of a particular constraint, or else we seek the result that minimises the weighted sum of the residuals for the original set of equations and the set of the residuals for the satisfaction of the constraints.

Absolute constraint   If the constraint is considered to be *absolute*, then the problem can be solved by the classical *Lagrange multiplier* technique. Thus the continuity constraint

$$\sum h_i = 1 \tag{3.19}$$

is a special case of the linear constraint

$$\underline{c}^T \underline{h} = b \tag{3.20}$$

with the special properties that c is a column vector with elements of unity and b is a scalar of value unity. To minimise the sum of squares of residuals given by equation (3.15) subject to the constraint of equation (3.20) it is necessary to minimise the new Lagrangian function

$$F(\underline{h}, \lambda) = r^T \cdot r + \lambda(b - \underline{c}^T \underline{h}) \tag{3.21}$$

Differentiating as before with respect to $\underline{h}$, we obtain as a modification of equation (3.18), the result:

$$(\underline{X}^T \underline{X}) \underline{h}_{opt} = \underline{X}^T \underline{y} - \frac{\lambda}{2} \underline{c} \tag{3.22}$$

In practice, the above equation is solved for a number of values of λ, and the particular value for which equation (3.20) holds is found by trial and error.

Desirable constraint   If the constraint requirement is *desirable* rather than absolute, then we seek to minimise the weighted sum of the residuals from the basic equation represented by (3.12) and the general constraint

$$\underline{C} \underline{h} = \underline{b} \tag{3.23}$$

where $\underline{C}$ is a matrix and $\underline{b}$ a vector. The function to be minimised is

$$F(\underline{h}, \gamma) = r^T \cdot r + \gamma (\underline{b} - \underline{C} \underline{h})^T (b - \underline{C} \underline{h}) \tag{3.24}$$

where γ is a weighting factor, which reflects the relative weight, given to the constraint conditions. The general solution to equation (3.24) is given by

$$(\underline{X}^T \underline{X} + \gamma \underline{C}^T \underline{C}) h_{opt} = \underline{X}^T \underline{y} + \gamma C^T b \tag{3.25}$$

In practice, the choice of γ is subjective and it is usually taken as the smallest value which eliminates the undesirable features of the *unconstrained least squares* solution.

In the case of *conceptual models*, the *parameters of the model* must be optimised in some sense. If we have chosen a specific model, then the predicted output is a function of the input and of the parameters of that model. Thus, in the case of a simple model with three parameters, we could write

$$\hat{y} = \phi[\underline{x}, a, b, c] \tag{3.26}$$

where $\underline{x}$ is the input, a, b, and c are the parameters of the model, and $\underline{y}$ is the output predicted by the model. The problem of optimisation is to find values of a, b and c so that the predicted values of $\underline{y}$ are as close as possible to the measured values of $\underline{y}$ in some sense to be defined.

The most common criterion is that the sum of the squares of the differences between the predicted outputs and the actual outputs will be a minimum (usually called the "*method of least squares*")

$$E_1(a, b, c) = \sum_i (y_i - \hat{y}_i)^2 = \text{min!} \tag{3.27}$$

As an alternative to using a least squares criterion, we could adopt the *Chebyshev criterion*

$$E_2(a, b, c) = \max_i |y_i - \hat{y}_i| = \text{min!} \tag{3.28}$$

i.e. minimise the maximum error.

Moment matching   Another criterion which can be used is *moment matching*. If we equate the first n statistical moments of the model and the prototype

$$\mu_R(\hat{y}) = \phi(a, b, c) = \mu_R(y) \tag{3.29}$$

the two systems are equivalent in that sense. When a large number of parameters are involved, the method of moment matching is not suitable, because higher order moments become unreliable due to the distorting effect of errors in the tail of the function, on the values of the moments. However, the method of moments has the great value that in cases where the moments of the model system can be expressed as a simple function of the parameters of the model, the parameters can be derived relatively easily.

For criteria such as *least squares* or *minimax error*, direct derivation of the optimum value of model parameters may be far from easy. In certain cases, it is possible to express the criterion to be minimised as a function of the parameters. We can differentiate this function with respect to each parameter in turn, set all the results equal to zero, and solve the resulting simultaneous equations to find the optimal value of the parameters. For any but the simplest model, it will probably be simpler to optimise the parameters by using a systematic search technique to find those parameter values, which give the minimum value of the error function.

The optimisation of model parameters by a *systematic search* technique is a powerful approach made possible by the use of digital computers. It is, however, not quite as easy as it might at first appear. In the almost trivial case of a two-parameter model, the problem of optimising these parameters subject to a least squares error criterion, can be easily illustrated. We can imagine the two parameters a and b as variables measured along two co-ordinate axes. The squares of the deviations between the predicted and actual outputs can be indicated by contours in the space defined by these axes. The problem of optimising our parameters is then equivalent to searching this relief map for the highest peak or the lowest valley, depending on the way in which we pose the problem. We have to search until we get, not merely a local optimum (maximum or minimum) but an absolute optimum. To examine every point of the space would be prohibitive, even in this simple example. In using a search technique, we have no guarantee that we will find the true optimum.

3.3 ORTHOGONAL FUNCTIONS

A set of functions

$$g_0(t), g_1(t), \ldots g_m(t) \ldots g_n(t) \ldots$$
$$\overline{g_0}(t), \overline{g_1}(t), \ldots \overline{g_m}(t) \ldots \overline{g_n}(t) \ldots$$

is said to be *orthogonal* on the interval $a < t < b$ with respect to the positive *weighting function* $w(t)$ if the functions satisfy the relationships:

$$\int_a^b w(t) g_m(t) \overline{g}_n(t) \, dt = 0, \quad m \neq n \tag{3.30}$$

$$\int_a^b w(t) g_m(t) \overline{g}_n(t) \, dt = \gamma_n, \quad m = n \tag{3.31}$$

where the standardisation factor (γ_n) is a constant depending only on the value of n. Equations (3.30) and (3.31) can be combined as follows

$$\int_a^b w(t)\,g_m(t)\,\overline{g}_n(t)\,\mathrm{d}t = \delta_{mn}\gamma_n \tag{3.32}$$

where δ_{mn} is the *Kronecker delta*, which is equal to unity when $m = n$, but zero otherwise.

Complete sets of orthogonal functions

If a function is expanded in terms of *complete* sets of orthogonal functions as defined above:

$$f(t) = \sum_{k=0}^{\infty} c_k g_k(t) \tag{3.33}$$

the property of orthogonality can be used to show that the coefficient (c_k) in the expansion in equation (3.33) is uniquely determined by

$$c_k = \frac{1}{\gamma_k} \int_a^b w(t)\,\overline{g}_n(t)\,f(t)\,\mathrm{d}t \tag{3.34}$$

If each of the functions $g_k(t)$ is so written that the factor of standardisation (γ_k) is incorporated into the function itself, the set of functions is said to be *orthonormal* i.e. normalised as well as orthogonal.

Fourier series

The most common set of orthogonal functions used in engineering mathematics is the *Fourier series*. The vast majority of single-valued functions used in engineering analysis and synthesis can be represented by an infinite expansion of the form

$$f(t) = \frac{1}{2}a_0 + \sum_{k=1}^{\infty} (a_k \cos(kt) + b_k \sin(kt)) \tag{3.35}$$

It can be shown that sines and cosines are orthogonal over a range of length 2π with respect to unity as a weighting function and with a standardisation factor equal to π . Accordingly we can write

$$\int_a^{a+2\pi} \cos(mt)\cos(nt)\,\mathrm{d}t = \pi\delta_{mn} \tag{3.36a}$$

$$\int_a^{a+2\pi} \sin(mt)\sin(nt)\,\mathrm{d}t = \pm\pi\delta_{mn} \tag{3.36b}$$

$$\int_a^{a+2\pi} \cos(mt)\sin(nt)\,\mathrm{d}t = 0 \tag{3.36c}$$

Because the terms of the Fourier series have the property of orthogonality, the coefficients a_k in equation (3.35) can be evaluated from

$$a_k = \frac{1}{\pi} \int_{-\pi}^{\pi} \cos(kt) f(t) \, dt \qquad (3.37a)$$

$$b_k = \frac{1}{\pi} \int_{-\pi}^{\pi} \sin(kt) f(t) \, dt \qquad (3.37b)$$

From a systems viewpoint, the significance of equation (3.35) is that the function is decomposed into a number of elementary signals, each of which is sinusoidal in form.

For mathematical manipulation, it is frequently more convenient to write the expansion given in equation (3.35) as a complex Fourier series:

$$f(t) = \sum_{k=-\infty}^{\infty} c_k \exp(ikt) \qquad (3.38)$$

For this exponential form of the Fourier series, the property of orthogonality is expressed as

$$\int_{a}^{a+2\pi} \exp[i(m-n)t] \, dt = 2\pi \delta_{mn} \qquad (3.39)$$

where δ_{mn} is again the Kronecker delta. We can determine the complex coefficients in equation (3.38) as

$$c_k = \frac{1}{2\pi} \int_{-\pi}^{\pi} \exp(-ikt) f(t) \, dt \qquad (3.40)$$

The relationship between the two sets of coefficients are given by

$$c_k = \tfrac{1}{2}(a_k - ib_k) \qquad (3.41a)$$

$$c_{-k} = \tfrac{1}{2}(a_k + ib_k) \qquad (3.41b)$$

If the function being expanded is a real function, the coefficients a_k and b_k in equations (3.35) and (3.37) are real, whereas the coefficients c_k in equations (3.38) and (3.40) are complex.

Three other cases of classical orthogonal polynomials are the *Legendre polynomials* which are orthogonal on a finite interval with respect to a unit

Laguerre polynomials

weighting function, the *Laguerre polynomials* which are orthogonal from zero to infinity with respect to the weighting function $\exp(-t)$, and the *Hermite polynomials* which are orthogonal on an interval from minus infinity to plus infinity with respect to a weighting function $\exp(-t^2)$. Thus the Laguerre polynomials have the property

$$\int_{0}^{\infty} \exp(-t) L_m(t) L_n(t) = \delta_{mn} \qquad (3.42)$$

and can be shown to have the explicit form

$$L_n(t) = \sum_{k=0}^{n} (-1)^k \binom{n}{k} \frac{t^k}{k!}$$ (3.43)

All of the above polynomials have the property that expansion in a finite series gives a least squares approximation to the function being fitted.

Since we are frequently concerned in hydrology with data defined only at discrete points, we are interested in polynomials and functions, which are orthogonal under summation, rather than under integration, as in the case of the above continuous functions. By analogy with equation (3.32) a set of *discrete functions* can be said to be orthogonal if

Discrete functions

$$\sum_{s=a}^{b} w(s) g_m(s) g_n(s) = \gamma_n \delta_{mn}$$ (3.44)

where s is a discrete variable. Since sines and cosines are orthogonal under summation as well as integration, the Fourier approach can be applied to a discrete set of equally spaced data.

The other classical orthogonal polynomials are not orthogonal under summation, but discrete analogues of them exist. The discrete analogue of the Laguerre polynomial is defined by

$$M_n(s) = \sum_{k=0}^{n} (-1)^k \binom{n}{k} \binom{s}{k}$$ (3.45)

Meixner polynomial

and is usually referred to as a *Meixner polynomial*. This function will be referred to below in connection with the black-box analysis of hydrological systems (Dooge and Garvey, 1978).

It is often convenient to incorporate the weighting function $w(t)$ as well as the factor of standardisation (γ_n) into an orthogonal polynomial and thus to form an orthonormal function which satisfies the relationship

$$\int_{a}^{b} f_m(t) f_n(t) \, dt = \delta_{mn}$$ (3.46)

or the corresponding discrete relationship

$$\sum_{s=a}^{b} f_m(s) f_n(s) = \delta_{mn}$$ (3.47)

For the case of Laguerre polynomials, as defined by equation (3.43) above, this results in the *Laguerre function*

$$f_n(t) = \exp\left(-\frac{t}{2}\right) \sum_{k=0}^{n} (-1)^k \binom{n}{k} \frac{t^k}{k!}$$ (3.48)

which satisfies equation (3.46) (Dooge, 1965). For the Meixner polynomial defined by equation (3.45), which is the discrete analogue of the Laguerre polynomial, the weighting function is $(1/2)^s$ and the factor of standardisation $(2)^{n+1}$ can be absorbed to give the *Meixner function* defined by

$$f_n(s) = \left(\frac{1}{2}\right)^{(s+n+1)/2} \sum_{k=0}^{n} (-1)^k \binom{n}{k} \binom{s}{k}$$

(3.49)

which satisfies equation (3.47).

3.4 APPLICATION TO SYSTEMS ANALYSIS

If an output from a system is specified as a number of equi-distant discrete points, it can be fitted exactly at these points by a finite Fourier series of the form

$$y(s) = \frac{A_0}{2} + \sum_{k=1}^{p} A_k \cos\left(k\frac{2\pi s}{n}\right) + \sum_{k=1}^{p} B_k \sin\left(k\frac{2\pi s}{n}\right)$$

(3.50)

where $n = 2p + 1$ is the number of data points. Since there are only n pieces of information, it is impossible to find more than n meaningful coefficients A_k and B_k for the data. Taking advantage of the fact that the sines and cosines are orthogonal under summation, the coefficients in the finite Fourier series given by equation (3.49) can be determined from

$$A_k = \frac{2}{n} \sum_{s=0}^{n-1} \cos\left(k\frac{2\pi s}{n}\right) y(s)$$

(3.51a)

$$B_k = \frac{2}{n} \sum_{s=0}^{n-1} \sin\left(k\frac{2\pi s}{n}\right) y(s)$$

(3.51b)

where k can take on the integral values $0, 1, 2, \ldots, p - 1, p$. The above formulation in equations (3.50) and (3.51) can also be expressed in the exponential form:

$$y(s) = \sum_{k=-p}^{p} C_k \exp\left(ik \cdot \frac{2\pi s}{n}\right)$$

(3.52a)

$$C_k = \frac{1}{n} \sum_{s=0}^{n-1} \exp\left(-ik \cdot \frac{2\pi s}{n}\right) y(s)$$

(3.52b)

In the Fourier analysis of systems, we seek the Fourier coefficients of the output (C_k) as a function of the Fourier coefficients of the input (c_k) and the Fourier coefficients of the pulse response (γ_k). This can be

Linkage relationship

done by substituting for $y(s)$ in equation (3.51), the right hand side of the *discrete convolution* of the pulse response and the input volumes, as given by equation (1.19) above. Then by reversing the order of summation and using the orthogonality relationship twice, it can be shown that we have the following *linkage relationship* between the three sets of Fourier coefficients

$$C_k = nc_k\gamma_k \tag{3.53}$$

For the expansion in the trigonometrical form of equation (3.50) rather than the exponential form of equation (3.52a) the *linkage equation* takes the form

$$A_k = \frac{n}{2}(a_k\alpha_k - b_k\beta_k) \tag{3.54a}$$

$$B_k = \frac{n}{2}(a_k\beta_k + b_k\alpha_k) \tag{3.54b}$$

Substituting the discrete analogues of equation (3.41), $c_k = (a_k - ib_k)/2$, $C_k = (A_k - iB_k)/2$ and $\gamma_k = (\alpha_k - i\beta_k)/2$, into equation (3.53) yields equations (3.54a) and (3.54b), which O'Donnell (1960) used in the first application of harmonic analysis to hydrological systems.

The *harmonic method* of analysis of linear time-invariant systems described above is an example of a *transform method of identification*. The observed inputs and outputs are transformed from the *time domain* to the *frequency domain*. The information originally given as values or ordinates in time is transformed into information concerning the coefficients of trigonometrical series. The number of coefficients is equal to the number of data points. The linkage equation in the transform domain given by equation (3.53) or equation (3.54) enables the harmonic coefficients of the pulse response to be found. Equations (3.54a and b) provide for each value of k, two simultaneous linear algebraic equations for the kth pair of harmonic coefficients of the unknown pulse response (α_k, β_k). An explicit solution is easily found in terms of the corresponding harmonic coefficients of the input (a_k, b_k) and output (A_k, B_k).

Inversion of the pulse response

The *inversion of the pulse response* back to the time domain is simple, since knowledge of the coefficients $(\alpha_k, \beta_k; k = 1, \ldots, n)$ enables the pulse response to be written as a finite Fourier series and the ordinates may be easily obtained. If the input and output data are given continuously in time, a similar analysis to the above, may be carried out using the ordinary continuous Fourier series. The same expressions are obtained for the linkage equations except that the base length of the periodic output (T) replaces the number of data points (n).

It has been suggested that for the case of *heavily damped systems with long memories*, a transform method based on Laguerre functions or Meixner functions would be more suitable than one based on trigonometrical functions (Dooge, 1965). In this book, there is scope to discuss only the discrete case, in which *Meixner functions* are used as the basis of system identification (Dooge and Garvey, 1978). The Meixner function of order n is defined in equation (3.49) above. The presence of the factor $(1/2)^s$ ensures that the tail of the function approximates to the form of an exponential decline and it is this feature of the function that suggests its use for heavily damped systems.

As in the case of harmonic analysis, the output data can be expressed in terms of Meixner functions as

$$y(s) = \sum_{n=0}^{N} A_n f_n(s) \tag{3.55}$$

where the Meixner functions are given by equation (3.49). The input and the unknown pulse response can also be expanded in terms of Meixner coefficients a_n and α_n. Because of the orthogonality property, the coefficients can be obtained by summation. For example, the coefficients of the output are

$$A_n = \sum_{s=0}^{\infty} f_n(s) y(s) \tag{3.56}$$

and similarly for the input a_n and the pulse response α_n.

Linkage equation

As in the case of harmonic analysis, a *linkage equation* can be found between the coefficients of the output A_p, the coefficients of the input a_n and the coefficients of the pulse response α_n. This linkage equation is not as simple in form as in the case of harmonic analysis being given by:

$$A_p = \sum_{k=0}^{p} (\sqrt{2} a_{p-k} - a_{p-k-1}) \alpha_k \tag{3.57}$$

The solution for the unknown coefficients of the pulse response involves the solution of a set of simultaneous linear algebraic equations, the coefficient matrix for which is a triangular matrix.

3.5 FOURIER AND LAPLACE TRANSFORMS

Fourier transform

The Fourier transform is particularly useful in the analysis of transient behaviour of stable systems. From one point of view, the *Fourier transform* may be looked on as a limiting form of the Fourier series. The ordinary Fourier series applies to functions that are periodic outside the interval of integration and consist of an infinite series in which each term refers to a definite discrete frequency. If the interval of integration is increased indefinitely, the series represented by equation (3.38) will be replaced by an integral with continuous frequency ω as follows

$$f(t) = \frac{1}{2\pi} \int_{-\infty}^{\infty} F(\omega) \exp(i\omega t) \, d\omega \tag{3.58}$$

and the expression for the coefficient given by equation (3.40) will be replaced by another integral

$$F(\omega) = \int_{-\infty}^{\infty} \exp(-i\omega t) f(t) \, dt \qquad (3.59)$$

In equations (3.58) and (3.59), as compared with equations (3.38) and (3.40), $F(\omega)$ corresponds to C_k, integration replaces summation in the equation for function expansion, and the standardisation constant 2π appears in the series equation rather than the coefficient equation. It would be equally permissible to introduce the standardisation constant 2π in equation (3.59) and omit it from equation (3.58), or even to introduce the square root of the factor into each of the equations.

Instead of looking on equation (3.58) as a limiting form of equation (3.38), it is possible to consider it simply as the equation defining the transformation of $f(t)$ from the time domain to the frequency domain. Equations (3.58) and (3.59) have the advantage that, unlike equations (3.38) and (3.40), they are not confined to periodic phenomena. This advantage, however, is offset by the fact that, whereas equation (3.38) enables us to evaluate the function to a high degree of accuracy by knowing the values of c_k, equation (3.58), which represents the inversion of the Fourier transform, is not, by any means, as easy to handle.

If the system we are examining is not stable, or if the functions involved do not fulfill certain other conditions, we take the transform not of $f(t)$ itself, but of $f(t) \exp(-st)$, where s is a complex number with a positive real part. Making this change, gives us the bilateral or two-sided *Laplace transform* of the function $f(t)$

Laplace transform

$$F_B(s) = \int_{-\infty}^{\infty} \exp(-st) f(t) \, dt \qquad (3.60a)$$

where

$$s = c + i\omega \qquad (3.60b)$$

As ordinarily used, the Laplace transform is only defined between zero and plus infinity, and virtually all tables are for this unilateral Laplace transform. In this form we have:

$$F(s) = \int_{0}^{\infty} \exp(-st) f(t) \, dt \qquad (3.61)$$

Expanding $\exp(-st)$ in definition (3.61) yields a power series in s with coefficients equal to the moments of $f(t)$ arranged in rank order with alternating signs. Differentiating R times and taking the limit as s tends to zero, picks out the moment about the origin of order R.

Equation (3.61) is the Laplace transform equivalent of equation (3.59) above. The Laplace transform can be inverted to give the original function in the same way as equation (3.58) by using the expression:

$$F(t) = \frac{1}{2\pi i} \int_{c-i\infty}^{c+i\infty} F(s)\exp(ts)\,ds \tag{3.62}$$

Again equation (3.62) is difficult to solve, but must be used unless the function $F(s)$ can be found in a set of Laplace transform tables. Numerical inversion of the Laplace transform is quite difficult and involves the use of orthogonal functions to represent the Laplace transform and the inversion of these functions term by term.

For a function known only at discrete points at interval D, the Laplace transform must be replaced by the *z-transform*. This can be written as

$$Z[f(nD)] = L\left[\sum_{n=0}^{\infty} f(t)\delta(t-nD)\right] \tag{3.63a}$$

$$= \sum_{n=0}^{\infty} f(nD)\exp(-snD) \tag{3.63b}$$

$$= \sum_{n=0}^{\infty} f(nD)z^{-n} \tag{3.63c}$$

where

$$z = \exp(sD) \tag{3.63d}$$

Discrete transform

This *discrete transform* has properties similar to the Laplace transform and these have been tabulated.

Fourier transforms, Laplace transforms and z-transforms are widely used in the analysis of non-periodic phenomena. The *linkage equations in the transform domain* can be derived for these transforms in the same way as for the orthogonal functions discussed already in Section 3.4. In each of these three cases, the operation of convolution in the time domain is transformed to multiplication in the transform domain, as in the case of the Fourier series.

Linkage equation for convolution

To illustrate the approach, we derive the linkage equation for convolution in the s-space of the Laplace transform, following Kreider et al. (1966, pp. 206–208). We apply the unilateral Laplace transform (3.61) to the convolution integral (1.15a) derived in Chapter 1 for the case of a lumped, linear, time-invariant, causal, fully-relaxed system. Hence

$$Y(s) = \int_{t=0}^{\infty} y(t) \exp(-st) \, dt \tag{3.64a}$$

$$= \int_{t=0}^{\infty} \exp(-st) \int_{\tau=0}^{t} x(\tau) h(t - \tau) \, d\tau \, dt \tag{3.64b}$$

$$= \int_{t=0}^{\infty} \int_{\tau=0}^{t} \exp(-st) x(\tau) h(t - \tau) \, d\tau \, dt \tag{3.64c}$$

where the double integration is carried out over the 45° wedge-shaped region of the $t\tau$ plane described by the inequalities

$$0 \leq \tau \leq t, \quad 0 \leq t < \infty \tag{3.65a}$$

But this region is also described by

$$\tau \leq t < \infty, \quad 0 \leq \tau \leq \infty \tag{3.65b}$$

and hence the iterated integral may be written in reverse order as

$$= \int_{\tau=0}^{\infty} \int_{t=\tau}^{\infty} \exp(-st) x(\tau) h(t - \tau) \, dt \, d\tau \tag{3.66a}$$

or

$$Y(s) = \int_{\tau=0}^{\infty} x(\tau) \int_{t=\tau}^{\infty} \exp(-st) h(t - \tau) \, dt \, d\tau \tag{3.66b}$$

$$= \int_{\tau=0}^{\infty} x(\tau) \exp(-s\tau) \int_{t=\tau}^{\infty} \exp[-s(t - \tau)] h(t - r) \, dt \, d\tau \tag{3.66c}$$

We now make the change of variable in the integral (3.66c) putting $u = t - \tau$, $du = -d\tau$; when $t = \tau$, $u = 0$; when $t = \infty$, $u = \infty$; hence

$$Y(s) = \int_{\tau=0}^{\infty} x(\tau) \exp(-s\tau) \int_{u=0}^{\infty} \exp(-su) h(u) \, du \, d\tau \tag{3.67a}$$

$$Y(s) = \int_{\tau=0}^{\infty} x(\tau) \exp(-s\tau) H(s) \, d\tau \tag{3.67b}$$

$$Y(s) = H(s) \int_{\tau=0}^{\infty} x(\tau) \exp(-s\tau) \, d\tau \tag{3.67c}$$

$$Y(s) = H(s) X(s) \tag{3.67d}$$

and the proof is complete. Convolution in t-space becomes multiplication in s-space!

From the definition (3.61), $Y(0)$ and $X(0)$ are the "areas" under the graphs of the functions $y(t)$ and $x(t)$ respectively. Hence, the system oper-

Conserves mass   ation conserves mass $[Y(0) = X(0)]$, if and only if, $H(0) = 1$ and the "area" under $h(t)$ is unity.

If we take s to be a complex variable with no real part, we see that the Fourier transforms of the input, the output and the pulse response, of a linear time-invariant system, will also be connected by

$$Y(\omega) = H(\omega)X(\omega) \tag{3.68}$$

where $Y(\omega)$ is the Fourier transform of the output, $H(\omega)$ is the Fourier transform of the unknown pulse response and $X(\omega)$ is the Fourier transform of the input. The difficulty of inverting the Fourier transform can be avoided in numerical computation by the use of the *Fast Fourier transform* technique (Brigham, 1974).

Fast Fourier transform

The Fourier transform is also of importance in systems theory because the Fourier transform of a given function can be used to generate the moments of that function. The moment of any function $f(t)$ about the origin is defined by

$$U'_R(f) = \int_{-\infty}^{\infty} f(t)\, t^R \, dt \tag{3.69}$$

The Fourier transform of the function $f(t)$ is given by

$$F(\omega) = \int_{-\infty}^{\infty} \exp(-i\omega t) f(t)\, dt \tag{3.70}$$

If this Fourier transform is differentiated R-times with respect to ω we obtain

$$\frac{d^R}{d\omega^R}[F(\omega)] = (-i)^R \int_{-\infty}^{\infty} \exp(-i\omega t)\, t^R f(t)\, dt \tag{3.71}$$

and if ω is now set equal to zero, we obtain

$$\frac{d^R}{d\omega^R}\left[F(\omega)\right]_{\omega=0} = (-i)^R \int_{-\infty}^{\infty} t^R f(t)\, dt \tag{3.72a}$$

$$= (-i)^R U'_R(f) \tag{3.72b}$$

The application of equation (3.72) to equation (3.68) allows us to find a *linkage relationship between the moments* of the output on the one hand and the moments of the input and the impulse response on the other hand, through the application of Leibnitz's Rule for the differentiation of a product. This relationship is

Linkage relationship between the moments

$$U'_R(y) = \sum_{j=0}^{R} \binom{R}{j} U_j(h)\, U'_{R-j}(x) \tag{3.73}$$

which is of great use in systems theory. It can be shown that the relationship between the *moments about the centres of area* of the input, the output and the impulse response also satisfy the relationship of equation (3.73). Nash (1959) first used this form of the moment relationship in hydrology.

Cumulants

An even simpler relationship can be found for the *cumulants*, which are also used by statisticians in order to characterise the shape of distributions (Kendall and Stuart, 1958). The cumulants may be defined as the functions generated from the logarithm of the Fourier transform in the same way as the moments are generated from the Fourier transform itself i.e.

$$k_R(f) = (i)^R \frac{d^R}{d\omega^R} \left[\log F(\omega) \right]_{\omega=0} \tag{3.74}$$

The multiplication of the Fourier transforms in the linkage equation (3.68) becomes an addition when logarithms of the Fourier transforms are taken. When both sides of the equation are differentiated R times and the limit to zero taken, we find

$$k_R(y) = k_R(h) + k_R(x) \tag{3.75}$$

Theorem of Cumulants

the *linkage equation for cumulants* (the *Theorem of Cumulants*).

When the "area" under the graph of $f(t)$ is 1, it can be shown from (3.72) and (3.74) that

- the *first cumulant* of $f(t)$ is the *first moment* of $f(t)$ *about the origin* (*i.e. the centre of "area" or temporal mean of the function f*),
- the *second cumulant* is the *second moment about the temporal mean*, and
- the *third cumulant* is the *third moment about the temporal mean*.

There is no simple relationship between higher order moments and cumulants. The requirement for unit area arises from the differentiation of log F. The resulting $1/F$ will be one at $\omega = 0$, so that $F(0) = 1$ and $f(t)$ has unit area. The equivalence between cumulants and moments only holds for $R = 1, 2$ or 3. For higher values of R the relationship becomes increasingly more complex. For example, the fourth cumulant is equal to the fourth moment about the mean minus three times the square of the second moment about the mean, and is referred to in statistics as excess kurtosis.

Accordingly, let us assume that the input and output functions, x and y, have been normalised to have unit "area". If the system is also conservative, the "area" under $h(t)$ is one, and we find a simple relationship of the same form as expression (3.75) for the moments of x, h and y — the first moments ($R = 1$) taken about the common time origin, and the second and third moments ($R = 2$ and 3) taken about the temporal means of each of the three functions. This is the *linkage equation for moments*,

Theorem of Moments

and its proof is the *Theorem of Moments*. These nine moments — three for x, three for h and three for y — are frequently sufficient to describe hydrological systems that are approximately linear.

In the case of discrete data it can be shown (though not so easily — using 3.63) that similar relationships exist for the *discrete moments* and *cumulants* of the input, the output and the pulse response.

3.6 DIFFERENTIAL EQUATIONS

Ordinary differential equations are differential equations in a single independent variable. If we are dealing with a system with *lumped inputs* and *outputs*, any differential equations describing that system will be equations only of time, and hence ordinary differential equations. Ordinary differential equations are classified in respect of their order and degree. The *order* of a differential equation is the order of the highest derivative present in the equation. The *degree* of the equation is the power to which the highest derivative is raised. A *linear equation* must of necessity be of the first degree, because otherwise there would be an essential non-linearity, and the principle of the superposition would not apply. In a linear differential equation all the derivatives in the equation must be to the first power and their coefficients must not be functions of the dependent variable. The general form of such an equation is

$$a_0(t)\frac{\mathrm{d}^n y}{\mathrm{d}t^n} + \cdots + a_{n-k}(t)\frac{\mathrm{d}^k y}{\mathrm{d}t^k} + \cdots + a_n(t)y = f(t) \tag{3.76}$$

If the coefficients are functions neither of y, nor of t, then we have an ordinary differential equation with *constant coefficients* given by

$$\frac{\mathrm{d}^n y}{\mathrm{d}t^n} + \cdots + a_{n-k}\frac{\mathrm{d}^k y}{\mathrm{d}t^k} + \cdots + a_n y = f(t) \tag{3.77}$$

Equation (3.76) could represent the operation of a lumped *linear system*, but for equation (3.77) to represent the operation of a system, the system would have to be lumped, linear and time-invariant.

If we have a lumped linear time-invariant system, the output is given by the convolution of the input with the pulse response in accordance with equation (1.10). If we now take *Laplace transforms* of these functions we obtain

$$Y(s) = H(s)X(s) \tag{3.78}$$

System function

Padé approximation

Suppose the Laplace transform of the *impulse response* $H(s)$, which is often referred to as the *system function*, can be represented by a *Padé approximation* (Ralston and Wilf, 1960, p. 13) or a *rational function* i.e. by the ratio of two polynomials, equation (3.78) becomes

$$Y(s) = \frac{P(s)}{Q(s)}X(s) \tag{3.79}$$

where $P(s)$ and $Q(s)$ are polynomials. For the system to be *stable* it is necessary for the order of the polynomial in the denominator to be equal or greater than the order of the polynomial in the numerator.

Suppose the order of the denominator polynomial $Q(s)$ is n, and that of the numerator polynomial $P(s)$ is m. It can be shown that the system will be represented by an ordinary differential equation of the type given in equation (3.77), whose order will be equal to the order of the polynomial $Q(s)$ and the right hand side of which, $f(t)$, will be given by the input $x(t)$ and its first m derivatives. In fact, if equation (3.77) is written in terms of the differential operator D, it will be given by

$$Q_n(D)\left[y(t)\right] = P_m(D)\left[x(t)\right] = f(t) \tag{3.80}$$

The solution of ordinary differential equations is dealt with in standard mathematical textbooks.

The first step in solving an equation such as (3.80) is to look at the *homogeneous equation*

Homogeneous equation

$$Q_n(D)\left[y(t)\right] = 0 \tag{3.81}$$

and to postpone the solution of the full non-homogeneous equation until a solution of the above homogeneous equation has been found. The classical method of solving equation (3.81) is to assume that the solution is made up of terms of an exponential form i.e.

$$y_k = c_k \exp\left(s_k t\right) \tag{3.82}$$

Any value of s which satisfies

$$Q_n(s) = 0 \tag{3.83}$$

will give a solution of equation (3.81). If the original equation is of the n-th order, there will be n roots s_k(real or complex) of equation (3.83). Consequently the general solution of equation (3.81), which is known as a *complementary function*, is given by

Complementary function

$$y = \sum_{k=1}^{n} c_k \exp\left(s_k t\right) \tag{3.84}$$

Real values of s_k give rise to exponential terms, and imaginary values of s_k to sinusoidal terms. In hydrological systems, which are *heavily damped,* the roots are usually negative and real, so that the solution consists of a series of exponentials with negative arguments. The n unknown constants c_k in equation (3.84) are obtained from the *boundary conditions*.

Having solved the homogeneous equation we then move on to the problem of solving the non-homogeneous equations. However, if a single particular solution of the *non-homogeneous equation* can be found

$$y = y_p(t) \tag{3.85}$$

Complete solution

then the *complete solution* of the non-homogeneous equation (3.80) is given by

$$y = y_p(t) + \sum_{k=1}^{n} c_k \exp\left(s_k t\right) \tag{3.86}$$

Particular integral

in which the first term or *particular integral* will satisfy the right hand side of equation (3.80) and the second term or complementary function will satisfy the boundary conditions.

The solutions of ordinary differential equations such as equation (3.80) can be greatly facilitated by the use of the *Laplace transform*. By taking the Laplace transform of the equation and using the rule for the Laplace transform of a derivative, we obtain an algebraic equation for the variables in which the boundary conditions are automatically incorporated. If the function on the right hand side of the equation is simple, its Laplace transform can be included. If not, it may be replaced by a delta function and the solution for this case obtained. The solution for the actual right hand side is then obtained by *convoluting* the function on the right hand side of the original equation with the solution obtained using a delta function.

This is equivalent to convoluting the input with the impulse response in order to obtain the output. If the right hand side of the equation involves derivatives, then the use of the delta function requires some caution and a mastery of its manipulation.

Partial differential equation

If the *system* has *distributed* rather than lumped characteristics then its operation will be described by a *partial differential equation*. Most of the partial differential equations encountered in engineering analysis are of the second order. For a single space dimension, the general second order homogeneous linear equation with constant coefficients is given by

$$a\frac{\partial^2 y}{\partial x^2} + b\frac{\partial^2 y}{\partial x \partial t} + c\frac{\partial^2 y}{\partial t^2} + d\frac{\partial y}{\partial x} + e\frac{\partial y}{\partial t} + fy = 0 \tag{3.87}$$

Partial differential equations are classified as hyperbolic, parabolic or elliptic according as the discriminant $b^2 - 4ac$ is respectively greater than, equal to or less than zero.

The equations governing unsteady flow with a free surface are hyperbolic in form, but are frequently simplified to a parabolic form in hydraulic and hydrological studies. The equations governing unsaturated flow through the soil and saturated flow of groundwater, are both parabolic in form. The full equations governing the above phenomena are non-linear and consequently their solution in their full form is difficult. Analytical solutions can only be obtained, if the equations are either linearised or greatly simplified.

3.7 REFERENCES ON SYSTEMS MATHEMATICS

This section is included to help those readers who find the mathematical concepts and techniques discussed above unfamiliar and difficult to understand. The books mentioned are ones which the authors have found to be clearly written and therefore suitable for someone relatively unfamiliar with the material. Those more expert in mathematics can easily choose from the books available in the library, those that may be more suitable for their level of attainment. The bibliographical details of the books referred to here, will be found in the alphabetic list of references at the end of the book.

Firstly it should be remembered that the topics discussed above will be found in standard reference books on applied mathematics, such as Korn and Korn (1961), Abramovitz and Stegun (1964) and Rektorys (1969). Books devoted to the mathematics of systems — such as Guillemin (1949), Brown (1965), Kreider et al. (1966), Raven (1966), Rosenbrock and Storey (1970) — will contain chapters on more than one of the topics covered in this chapter. Consequently a single book from this list may suffice for any extra reading necessary.

The treatment of matrices in mathematical textbooks varies widely in respect of viewpoint adopted. Anyone whose knowledge of the properties and manipulation of matrices is rudimentary would be well advised to consult a book whose aim is to be clear rather than compact and which contains suitable examples. Books, which can be recommended from this viewpoint, are Guillemin (1949), Bickley and Thompson (1964), Frazer, Duncan and Collar (1965) and Sawyer (1972). The basic computational aspects of matrix manipulation are dealt with in an introductory fashion by Hamming (1971), and in more detail by Demidovich and Maron (1973). Computer programming for matrix problems is discussed in Ralston and Wilf (1960, 1967) and in Carnahan, Luther and Wilkes (1969).

Optimisation, both unconstrained and constrained, is covered in many works from a number of different points of view. A good approach might be to read Chapter 9 of Hamming (1971) as an introduction. This could be followed by Wilde (1964) on search techniques, and by a linear programming text such as Dantzig (1963) or Gass (1964) as an introduction to the case where the elements of the solution must be non-negative.

Fourier series are described in many standard reference books including most of the general references cited in the second paragraph of this section. Other books containing good and readable accounts of the properties, use and computation of Fourier series are Guillemin (1949), Lanczos (1957), Ralston and Wilf (1960), Hamming (1962, 1971), Acton (1970) and Smith (1975). The classical orthogonal polynomials are dealt with in many general texts such as Abramovitz and Stegun (1964), and in texts on special functions, such as those by Rainville (1960) and Lebedev (1972). The application of orthogonal functions to hydrological systems has been discussed by O'Donnell (1960) in respect of harmonic analysis, by Dooge (1965) in respect of Laguerre polynomials, and by Dooge and Garvey (1977) in respect of Meixner polynomials.

Fourier transforms, Laplace transforms and z-transforms are also dealt with in many standard texts on mathematics and on systems methods. There are chapters on these topics in the books devoted to systems mathematics, which are listed in the second paragraph of this section. Other texts with good treatment of these topics are Aseltine (1958), Doetsch (1961), Papoulis (1962), Jury (1964), Kreider et al. (1966), Saucedo and Schiring (1968) and Brigham (1974). Extensive tables of Laplace transforms are available in Erdelyi (1954) and Roberts and Kaufman (1966).

Differential equations have also been treated in a vast number of texts from many points of view. Further details can be found in texts such as Sneddon (1957), Lambe and Tranter (1961), Fox (1962), Collatz (1966) and Petrovskii (1967).

CHAPTER 4

Black-Box Analysis of Direct Storm Runoff

4.1 THE PROBLEM OF SYSTEM IDENTIFICATION

The *black-box approach* to the analysis of systems has already been dealt with in outline in Chapter 1. It is based on the concept of system operation shown in Figure 1.1. The basic problem of black-box identification is the determination and mathematical description of the system operation on the basis of records of related inputs and outputs. In that chapter, it was pointed out that for the case of a lumped linear time-invariant system with continuous inputs and outputs, this mathematical description is given by the impulse response of the system. This contains all the information required for the prediction of the operation of such a system on other inputs. In the case of a linear lumped time-invariant system in which the input and output data are given in discrete form, the operation of the system can be

Pulse response mathematically described in terms of the pulse response. Of necessity, the pulse response contains less information than the impulse response. But it is adequate for predicting for any given input the corresponding output at discrete values of the sampling interval.

Accordingly, as pointed out in Section 1.4, the problem of the identification of a lumped time-invariant system by black-box analysis amounts to a solution of the set of linear algebraic equations

$$\underline{y} = \underline{X}\,\underline{h} \tag{4.1}$$

where $\underline{h}$ is the vector of unknown ordinates of the pulse response, $\underline{y}$ is the vector of the known ordinates of the output and $\underline{X}$ is the matrix formed as follows from the vector of the known input ordinates.

$$
\underline{X} =
\begin{bmatrix}
X_0 & 0 & \cdot & \cdot & \cdot & 0 & 0 \\
X_1 & X_0 & \cdot & \cdot & \cdot & 0 & 0 \\
\cdot & \cdot & \cdot & \cdot & \cdot & \cdot & \cdot \\
\cdot & \cdot & \cdot & \cdot & \cdot & \cdot & \cdot \\
X_m & X_{m-1} & \cdot & \cdot & \cdot & \cdot & \cdot \\
0 & X_m & \cdot & \cdot & \cdot & X_0 & 0 \\
0 & 0 & X_m & \cdot & \cdot & X_1 & X_0 \\
\cdot & \cdot & 0 & \cdot & \cdot & X_2 & X_1 \\
\cdot & \cdot & \cdot & \cdot & \cdot & \cdot & \cdot \\
\cdot & \cdot & \cdot & \cdot & \cdot & X_m & X_{m-1} \\
0 & 0 & \cdot & \cdot & \cdot & 0 & X_m
\end{bmatrix}
\tag{4.2}
$$

The basis for equation (4.2) is given in Chapter 1, where they appear as equations (1.20) and (1.21) respectively.

The *unit hydrograph approach* that is described in Chapter 2 of this book has been seen to be based on the assumption that the catchment converts effective rainfall to direct storm runoff in a lumped linear time-invariant fashion. The *finite-period unit hydrograph* is then seen to correspond to the *pulse response* of systems analysis and the *instantaneous unit hydrograph* to the *impulse response*. As mentioned in Chapter 2, in the early years unit hydrographs were derived either by trial and error (graphical or numerical) or by the solution of the linear equation by forward substitution. In obtaining these derived unit hydrographs by hand, any obvious errors were adjusted subjectively according to preconceived ideas concerning a realistic shape for the unit hydrograph. Later on, attempts were made to derive objective methods of unit hydrograph derivation, which could be applied to complex storm records and automated for the digital computer. Some of these approaches were briefly mentioned in Chapter 3 on Systems Mathematics.

Inversion process

Since the derivation of the unit hydrograph is essentially an *inversion process*, the effects of error in the data may appear in a magnified form in the derived unit hydrograph. It is always possible to derive an apparent unit hydrograph from a record of effective precipitation and direct storm runoff. Unless the method of derivation is grossly unsuitable or inaccurate, the recorded output can be approximated closely by the reconstructed output obtained by convoluting the recorded input and this estimated unit hydrograph. Unfortunately, however, the degree of correspondence between the predicted and the recorded output may, as a result of errors in the data, be a poor indicator of the correspondence of the estimated unit hydrograph to the "true" pulse response. Consequently, the ability of the estimated unit hydrograph to predict the direct storm response for a given pattern of effective precipitation, different to that from which it is derived, cannot be judged on the basis of the ability of the derived unit hydrograph to reproduce the output, and cannot be judged on the basis of the input and output data alone. The effect of data errors on unit hydrograph derivation can be studied systematically, either by a mathematical analysis of the techniques used for unit hydrograph derivation, or by numerical experimentation.

The suitability of any proposed method of system identification for application to real data is best evaluated by the adoption of a three-stage strategy.

In the *first stage*:

Synthetic set of input and output data

the validity of the proposed identification method is verified by applying it to a *synthetic set of input and output data* generated by choosing a specific system response and convoluting this with a chosen input in order to generate the corresponding synthetic output. The impulse response or the pulse response may then be estimated by applying the proposed identification method to the synthetic input and output. We now compare the derived system response to the known system response used in the generation of the output. If the variation of system response is appreciable, this indicates either some basic defect in the proposed method, or some error in applying the method, or an undue amount of round-off error, in either the generation of the data or the use of the method.

The *second stage*:

consists of the verification of the robustness of the method of system identification (which has been found to be valid in the first step) by examining the effect on the results of errors in the input and the output. Adding to "error-free" input and output data, error of a known type and magnitude, allows us to test the ability of the method to derive an acceptable approximation to the true unit hydrograph, or other response, in the presence of such error.

The *third stage*:

Only after the validity and robustness of the method of system identification has been verified, as described above, may the method be applied safely to the linear analysis of actual field data.

The approach may be summarised in the following procedure.

Step		Given	Calculate	Remarks
1.	Make a perfect data set	$x(t), h(t)$	$y(t)$	Solve the problem of system prediction
2.	Make a working data set	$\varepsilon(x), \varepsilon(y)$	$x' = x + \varepsilon(x)$ $y' = y + \varepsilon(y)$	Corrupt the data with systematic and random error of size ε
3.	Test each method of system identification	$x'(t), y'(t)$	$h'(t)$	Estimate the "true" unit hydrograph with each method
4.	Measure the performance of each method	$\varepsilon(x), \varepsilon(y)$ $h(t), h'(t)$	$\lVert \varepsilon \rVert,$ $\lVert h(t) - h'(t) \rVert$	Calculate vector norms for the error and the corresponding impulse response

Effects of errors

Laurenson and O'Donnell (1969) carried out the first comprehensive study on the effects of errors on unit hydrograph derivation.

4.2 OUTLINE OF NUMERICAL EXPERIMENTATION

The remainder of this chapter will be devoted to an outline of the pioneering study by Laurenson and O'Donnell (1969) and of its extension by the senior author and two of his postgraduate students (Garvey, 1972; Bruen, 1977; Dooge, 1977, 1979). In their study Laurenson and O'Donnell assumed the impulse or instantaneous unit hydrograph to be

$$h(t) = \left[\left(\frac{1}{t+T} - \frac{1}{20+T} \right) \exp \left(-\frac{20}{At+T} \right) \right]^R \tag{4.3}$$

for all values of t between 0 and 20 and to be zero outside these limits. The model represented by equation (4.3) has three parameters and could

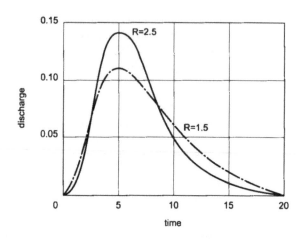

Figure 4.1. Shape of unit
hydrograph in numerical
experimentation.

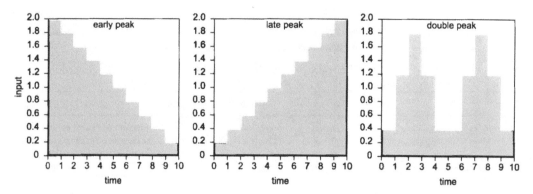

Figure 4.2. Input shapes.

be used to generate a wide variety of shapes of response. In their study,
Laurenson and O'Donnell experimented with only two sets of the values
of the parameters T, A and R. These two shapes are shown in Figure 4.1. In
each case $T = 2.5$ and $A = 3.0$. In the case of the thinner unit hydrograph
$R = 2.5$ and in the case of the fatter unit hydrograph $R = 1.5$. In the original
study by Laurenson and O'Donnell (1969), three shapes of rainfall input
were used. These are shown in Figure 4.2.

The combination of the three alternative input patterns shown in
Figure 4.2 with the two alternative shapes of instantaneous unit hydrographs shown in Figure 4.1 gave rise to six sets of synthetic input–output
data. This synthetic error-free data was then contaminated by systematic error of the type and magnitude likely to occur in hydrological
measurements.

Types of systematic
error

The following six *types of systematic error* were studied:

(1) volume of total rainfall;

(2) rainfall synchronisation;
(3) rainfall-runoff synchronisation;
(4) rating curve for discharge;
(5) base flow separation;
(6) assumption of uniform loss rate.

The effects of the above six types of error were studied for three methods of black-box analysis (least squares, harmonic analysis and Meixner analysis) and one method based on a conceptual model (cascade of equal linear reservoirs).

Types of
random error

The study of Laurenson and O'Donnell (1969) was extended by Garvey (1972) who tested nine methods of black-box analysis and three conceptual models for their stability in the presence of six *types of random error* as well as the six types of systematic error previously studied. These six types of random error were

(1) error in input only and proportional to the maximum ordinate;
(2) error in input only and proportional to individual ordinates;
(3) error in output only and proportional to maximum ordinate;
(4) error in output only and proportional to individual ordinates;
(5) error divided equally between input and output and proportional to maximum ordinates;
(6) error divided equally between input and output and proportional to individual ordinates.

Garvey (1972) also investigated the effect of the shape of the unit hydrograph on the fitting of conceptual models by using seven sets of parameters in the unit hydrograph equation given by equation (4.3). He also studied the effect of three different levels (5%, 10%, 15%) of error in the data on the mean error in the unit hydrograph (Garvey, 1972).

Bruen (1977) later extended Garvey's investigation and his computer program, by increasing the number of inputs studied from 3 to 6, the number of methods of black-box analysis from 9 to 15, the number of conceptual models from 3 to 25, the number of types of random error from 6 to 12 and the number of levels of error studied from 3 to 6. He also extended the study to compute the mean and variance of a large number of *realisations* for each case of random error rather than the individual realisations of the random process, which was done by Garvey (1972). Another point studied in Bruen's project of exploratory computation was the effect of "filtering" either the input–output data (i.e. pre-filtering), or the estimated unit hydrograph (i.e. post-filtering). A *filter* in this sense is an operation, which removes or reduces an unwanted characteristic in the record. Thus the truncation of the Fourier series representation of a function, or of a data series, removes contributions from frequencies above the cut-off frequency, and is a numerical frequency filter equivalent to an ideal low-pass filter with the same cut-off frequency. The most important filters examined by Bruen were

(1) smoothing the derived unit hydrograph by a *moving average* filter in the time domain or by cut-off filter in the frequency domain (filter type S);
(2) maintaining *non-negativity* of the ordinates of the unit hydrograph by setting all negative ordinates equal to zero (filter N);
(3) imposing a *mass continuity* condition on the ordinates of the unit hydrograph by normalising the sum of the ordinates (filter A).

4.3 DIRECT ALGEBRAIC METHODS OF IDENTIFICATION

One obvious approach to the solution of the problem of identification for a lumped linear time-invariant system is to solve for the unknown values of the unit hydrograph vector by direct algebraic solution of the set of linear equations described by equation (4.1). The number of equations in the system is determined by the number of ordinates in the output vector $(y_0, y_1, \ldots, y_{p-1}, y_p)$. If the number of unknown ordinates of the unit hydrograph is taken equal to the number of output ordinates then the matrix X in equation (4.1) is a square matrix and if it is non-singular, it can be inverted. However, consideration of the definition of the finite unit hydrograph indicates that the number of ordinates in the output $(p + 1)$, the number of ordinates in the input $(m + 1)$ and the number of ordinates in the unit hydrograph $(n + 1)$ are connected through the relationship:

$$p = m + n \tag{4.4}$$

Accordingly the *number of ordinates* in the unit hydrograph $(n + 1)$ will be less the number of ordinates in the output $(p + 1)$ and we can write

$$h_i = 0 \quad \text{for } i > (n + 1) \tag{4.5}$$

The elimination of these values of h_i involves the elimination of the corresponding columns of the input matrix X thus reducing it from a $(p + 1, p + 1)$ matrix to a $(p + 1, n + 1)$ matrix.

The reduced form of matrix X obtained by making the assumption of equation (4.5), can be solved by direct matrix inversion by choosing any $(n + 1)$ of the rows and inverting the resulting square matrix. It can be seen from equation (4.2) that if the first $(n + 1)$ rows are chosen then the matrix to be inverted will be lower triangular and can be solved directly

by *forward substitution*. The solution for any step is given by

$$h_i = \frac{y_i - \sum_{j=0}^{i-1} x_{i-j} h_j}{x_0} \tag{4.6}$$

which can be solved iteratively for all values of i from $i = 0$ to $i = n + 1$. Similarly, if the last $(n + 1)$ equations are taken, then the matrix to be inverted is upper triangular and the problem can be solved by *backward substitution*.

When the set of equations are solved by forward substitution the results are found to be extremely sensitive to the shape of the input and also to the presence or absence of post-filtering. Table 4.1 summarises some of the results for forward substitution obtained by Garvey (1972). The table shows the mean absolute error in the derived unit hydrograph as a percentage of the true peak value for (a) synthetic error-free data (mean of six input–output cases), (b) input–output data with 10% systematic error (mean of 36 cases), and (c) input–output data with 10% random error (mean of 36 cases).

The first line in the table shows that for the error-free case there is a very small error in the derived unit hydrograph due to roundoff error in the computation. For the case of 10% error, either systematic or random, there is a complete numerical explosion and the results are worthless. If the constraint is applied that all negative ordinates are set equal to zero (i.e. a non-negativity or type N filter) there is no improvement in the situation. If, however, the constraint of unit area (a type A filter) is imposed then the results are no longer completely explosive though still highly inaccurate. As can be seen from Table 4.1, in the latter case the error in the derived unit hydrograph for 10% error in the data is 252% for systematic error and 964% for random error. When both constraints are applied these errors are reduced to 27% and 46% respectively. In the latter case where both filters are applied, though the numerical stability has been brought under control, the accuracy of the results is still not acceptable for practical purposes.

It is clear from equation (4.6) that the propagation of any error that arises in an ordinate of the derived unit hydrograph will be affected by the value of x_0. Accordingly one would expect different results to be obtained for the *early peaked and late peaked rainfall* patterns shown in Figure 4.2. For the case of no constraint, the early-peaked pattern of input in which x_0 is the highest value of input, shows no sign of a numerical explosion whereas for the other two patterns of input there is such an explosion for

Table 4.1. Effects of constraints on forward substitution solution.

Constraint	Mean absolute error as % of peak		
	Error-free	Systematic error	Random error
None	0.9×10^{-3}	1×10^{9}	2×10^{9}
Non-negativity	0.9×10^{-3}	1×10^{11}	2×10^{9}
Normalised area	0.9×10^{-3}	252	964
Both constraints	0.9×10^{-3}	27	46

all six cases of random error and for the two systematic cases of error in the total rainfall or a systematic error in synchronisation between the rain gauges. If, in fact, backward substitution were used instead of forward substitution the late-peaked input would be found to be stable and the early-peaked to be highly unstable. Thus neither method is stable for all shapes of precipitation input. When the constraints of unit area and non-negativity are applied the differences are less marked, but are nevertheless significant as shown by Table 4.2.

Collins procedure

This problem of sensitivity to input shape can be overcome to some degree by adopting the *procedure proposed by Collins* (1939) which is referred to in Section 2.3. The iterative computation suggested by Collins can be adapted for the computer by making explicit the assumption implicit in the iterative method that the system is causal and hence that all ordinates of the derived unit hydrograph for negative time are ignored. In matrix terms the method consists essentially of ignoring all equations which do not contain the maximum input ordinate i.e. of solving the $(n + 1)$ equations starting with the first equation which contains the maximum input ordinate. In this way the $\underline{X}$ matrix is reduced to a square matrix and can be inverted. The choice of this particular set of equations ensures that the diagonal elements of the matrix to be inverted are greater than the off-diagonal elements, which improves the stability of the matrix inversion.

A comparison of the results of the Collins method with those for forward substitution and backward substitution is shown in Table 4.3, which is based on numerical experiments by Garvey (1972). It shows that the

Table 4.2. Effects of rainfall pattern on error in unit hydrograph.

Input pattern	Mean absolute error as % of peak		
	Error-free	Systematic error	Random error
Early-peaked	0.72×10^{-3}	14.3	20.3
Late-peaked	1.00×10^{-3}	40.7	57.1
Double-peaked	0.84×10^{-3}	26.0	59.2
Average	0.85×10^{-3}	27.0	45.5

Table 4.3. Comparison of direct algebraic solutions.

Methods used	Mean absolute error as % of peak		
	Error-free	Systematic error	Random error
Forward substitution	0.9×10^{-3}	27.0	45.5
Backward substitution	0.4×10^{-3}	28.0	35.5
Collins method	0.1×10^{-3}	5.8	27.7

Collins method is much more effective in reducing the error in the derived unit hydrograph for the case of systematic error compared with random error.

The result for systematic error could be considered satisfactory, since the error in the derived unit hydrograph is substantially below the 10% level of error in the data. However, the result for random error indicates that the Collins method would not be satisfactory, if the level of random error were of the order of 10%. Nevertheless, it may be concluded that, if a direct algebraic method is to be used, the Collins method should be chosen.

4.4 OPTIMISATION METHODS OF UNIT HYDROGRAPH DERIVATION

Method of least squares

The obvious starting point for any discussion of an optimisation approach to the problem of unit hydrograph derivation is the *method of least squares*, which was discussed in Chapter 3. While the methods of solution described in the last section seek to satisfy exactly some chosen $(n + 1)$ set of the available $(p + 1)$ equations, the least squares method seeks a solution that will be a best fit to all $(p + 1)$ equations. By best fit is meant the solution which minimises the sum of the squares of the differences between the predicted and measured outputs. This will certainly give a *smoother approximation* to the whole range of output, but what we are concerned with, is whether it will give a better approximation to the true system response. It was shown in Chapter 3 that the least squares estimate of the pulse response h can be obtained by the solution of the set of equations

$$(\underline{X}^T \underline{X})h = \underline{X}^T \underline{y} \tag{4.7}$$

through the inversion of the square matrix $(\underline{X}^T \underline{X})$. Whether this will give an improved method of solution to the problem of unit hydrograph derivation depends on whether the latter matrix is better conditioned than the original matrix from which it was derived. The least squares method was applied to unit hydrograph derivation by Snyder (1955) and Body (1959), and later improved by Newton and Vinyard (1967) and by Bruen and Dooge (1984). See also Dooge and Bruen (1989).

The numerical experiments by Garvey (1972) indicated that the results for the least squares method were substantially independent of the *shape of input pattern* as summarised in Table 4.4.

Comparing the last lines of Table 4.3 and Table 4.4, it can be seen, that for the case of random error, the least squares method gives slightly better results than the Collins method, but the performance for random error is still unsatisfactory.

It is interesting to examine the effect of the *type of random error* in the data on the error in the derived unit hydrograph. It will be recalled that in the experiments by Garvey (1972) there were six types of random error, two based on errors on the input only, two based on errors in the output only and two based on errors equally divided between the input and the output. When Garvey's results are classified according to type of error as in Table 4.5, we see that the unsatisfactory performance of the least squares method, occurs when either all or part of the *error is in the output*.

The differences shown in Table 4.5 are easily explained, if we consider carefully what has been done. The least squares method is based essentially on the attempt to match the output as closely as possible. If the output is in error, the method will seek to match the given output including the error in that output. The attempt to match the error as well as the true output, results in errors in the derived unit hydrograph. In the case where there are errors in the input, the errors in the derived unit hydrograph are less, because the output being fitted is correct.

More recently, further developments in the method of least squares have been applied to the identification of hydrological systems. These involve the incorporation into the actual inversion procedure itself, of *constraints* such as normalisation of the area, or non-negativity of ordin-

Post-filters   ates, or a smoothing constraint. These are applied as *post-filters* to the direct algebraic methods, or to the unconstrained least squares solution described above. The relationship between the various optimisation procedures, which operate on the deviations between predicted and observed outputs, are shown in Figure 4.3.

Table 4.4. Effects of input pattern on least squares solution.

Input pattern	Mean absolute error as % of peak		
	Error-free	Systematic error	Random error
Early-peaked	0.2×10^{-3}	6.7	20.7
Late-peaked	0.2×10^{-3}	6.5	19.4
Double-peaked	8.5×10^{-3}	6.7	24.3
Average	3.0×10^{-3}	6.6	21.5

Table 4.5. Effects of error type on least squares solution.

Type of error	Mean absolute error as % of peak			
	Early-peaked	Late-peaked	Double-peaked	Average
Input only	2.7	6.6	3.3	4.2
Output only	34.4	30.2	39.9	34.8
Input and output	25.5	21.5	29.7	25.4
Average	20.7	19.4	24.3	21.5

The *method of regularisation* was applied to hydrology by Kuchment (1967) and that of *quadratic programming* by Natale and Todini (1973). If the objective function is taken as the minimisation of the absolute deviation of the predicted output from the observed output, the problem can be formulated as one of *linear programming*. Deininger (1969) applied this approach to hydrological systems. The numerical experiments by Garvey (1972) indicated that the methods of regularisation could reduce the errors in the derived unit hydrograph in the presence of random error, but that no improvement was obtained by the use of linear programming. A summary of the results is shown in Table 4.6 together with comparative times of computation.

It can be seen from Table 4.6 that the method of regularisation reduces the mean error in the derived unit hydrograph, for the six cases of random error, to a level comparable to the error in the original data, but at the cost of increased time of computation.

The problem of the propagation of error in matrix substitution methods *Condition number* may be analysed by the use of the condition number. The least squares

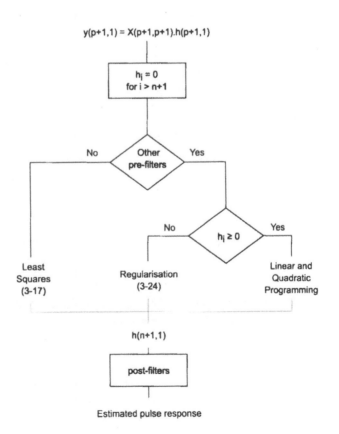

Figure 4.3. The relationship between optimisation methods.

Table 4.6. Comparison of optimisation methods.

Method used	Mean absolute error as % of peak			Relative CPU time
	Error-free	Systematic error	Random error	
Least squares	0.3×10^{-3}	6.6	21.5	1
Regularisation	3.1×10^{-3}	4.3	11.5	5.8
Linear programming	480×10^{-3}	11.6	23.1	9.7

solution is obtained by inverting the multiplier of the unit hydrograph on the left hand side of equation (4.7) above to obtain

$$h_{opt} = (X^T X)^{-1} X^T y \qquad (4.8)$$

The quantity

$$K(x) = \|X^{-1}\| \cdot \|X\| \qquad (4.9)$$

is defined as the condition member of X where $\| \cdots \|$ denotes one of several possible kinds of matrix norm. It can be shown that the condition number provides an upper bound for the magnification of error in the inversion process of equation (4.8). It can be shown that the lowest of

Euclidean vector norm   these upper bounds is given by the Euclidean vector norm

$$\|X\|_2 = \left[\sum_{k=1}^{m} \sum_{j=1}^{m} (x_{jk})^2 \right]^{1/2} \qquad (4.10)$$

and the special matrix norm induced by this vector norm. For this case, the condition number defined by equation (4.9) is given by

$$K(x) = \left(\frac{\lambda_{max}}{\lambda_{min}} \right)^{1/2} \qquad (4.11)$$

where λ_{max} and λ_{min} are the maximum and minimum eigenvalues of the Toeplitz matrix $X^T X$.

The condition number for the algebraic methods of Section 4.3 (forward substitution, backward substitution, Collins method) and for the optimisation methods of Section 4.4 above (least squares, regularisation, pre-whitening) for the simplistic numerical example of two unknown unit hydrograph ordinates, involves only the solution of a quadratic equation (Dooge and Bruen, 1989). A comparison of these methods based on analytically derived condition numbers, indicates that the methods rank as in Tables 4.3 and 4.6 above and provides a clear explanation of the results obtained in the numerical experiments described above.

As indicated in Tables 4.3 and 4.6 above, the best results were obtained by smoothed least squares (Bruen and Dooge, 1984) which is a special case of regularisation (Kuchment, 1967). This approach was subsequently

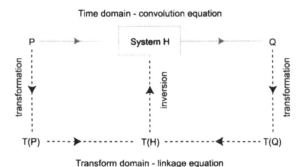

Figure 4.4. The procedure in transform methods.

extended from an *a priori* assumption of smoothness in the unit hydrograph to an *a priori* assumption of a smooth unit hydrograph shape based on regional catchment information (Bruen and Dooge, 1992a, 1992b).

4.5 UNIT HYDROGRAPH DERIVATION THROUGH Z-TRANSFORMS

Both the methods based on direct matrix inversion and those based on optimisation discussed above, seek a solution to the basic problem of unit hydrograph derivation by determining the elements of the vector of unknown unit hydrograph ordinates in the time domain. An alternative is to seek some other representation for the three functions involved, which may be more convenient for solution purposes. This general procedure is shown in outline in Figure 4.4. It involves

(a) the transformation of the three functions of input/output and pulse response to the transform domain;
(b) the formulation of the linkage equation in that transform domain;
(c) the solution of the linkage equation in order to obtain the transform of the pulse response (i.e. the finite period unit hydrograph), and finally
(d) the inversion of the derived pulse response from the transform domain to the time domain.

Transform methods

In choosing between *transform methods*, regard must be had for

(a) the relative simplicity or difficulty of the three operations of transformation, linkage solution and inversion; and
(b) whether the type of basis function involved in the transformation is a suitable one for the type of system under study.

The fact that Laplace transform methods are widely used in the analysis of linear systems in Electrical Engineering, suggests that the problem of discrete convolution represented by equation (4.1) should be tackled by means of the corresponding z-transform defined by:

$$F(z^{-1}) = \sum_{s=0}^{\infty} f(s) z^{-1} \qquad (4.12)$$

where the notation $F(z^{-1})$ is used instead of the more usual $F(z)$ in order to emphasise the fact that the *z-transform* is a polynomial in z^{-1} rather than in z. For this transformation (as for the Laplace transform) the operation of convolution in the time domain, corresponds to the operation of multiplication in the transform domain. The relationship between input and output of a lumped linear time-invariant system whose operation is defined by equation (4.1) above, is given in the transform domain by

$$Y(z^{-1}) = X(z^{-1}) H(z^{-1}) \qquad (4.13)$$

where $H(z^{-1})$ is the z-transform of the pulse response and is often termed the *discrete system function*.

The solution of the linkage equation for (4.13) is obviously given by

$$H(z^{-1}) = \frac{Y(z^{-1})}{X(z^{-1})} \qquad (4.14)$$

If a robust method of determining $H(z^{-1})$ from this equation is available, the inversion to the discrete time domain offers no difficulty. The ordinate of the pulse response or unit hydrograph $h(s)$ for any value of s will be given by the coefficient *of* z^{-s} in the system function $H(z^{-1})$.

Polynomial division

At first sight one would be inclined to proceed to the solution of equation (4.14) by the direct *polynomial division* of $Y(z^{-1})$ by $X(z^{-1})$ thus obtaining the coefficients of z^{-s} in $H(z^{-1})$. This straightforward method will give acceptable results for error-free data, but will prove most unstable in the presence of errors in the data. In fact, it is quite easy to show that the solution of equation (4.14) by polynomial division is equivalent in every way to the solution of the set of equations given by equation (4.1) by the method of forward substitution already discussed under the heading of direct algebraic methods.

If, however, the z-transform polynomials are specified, not in terms of their coefficients, but in terms of their roots, the prospect for a robust method of solution becomes somewhat brighter. De Laine (1970) sought to derive the unit hydrograph in the absence of rainfall data by using the output data for a series of storms on the same catchment. He argued that

(a) the roots of the output polynomials which recur in every storm, must belong to the system function polynomial rather than to the unknown input polynomials; and

(b) these common roots can therefore be used to reconstitute the unit hydrographs in the time domain.

Table 4.7. Root-matching solution for 10% random error.

Type of error	Mean absolute error as % of peak			Average for three shapes
	Early-peaked	Late-peaked	Double-peaked	
Input only	1.9	1.5	1.6	1.7
Output only	33.4	30.8	41.9	35.3
Equal error	25.8	28.1	29.0	27.6
Average error	20.4	20.1	24.1	21.5

Matching the roots

De Laine's approach can be adapted for use in the case of a single input/output event by *matching the roots* of the input polynomial with the roots closest to them in the output polynomial, and reconstituting the pulse response of the system of the time domain from the remaining roots. In the presence of data errors, the polynomial roots will change in value and the matching of the roots becomes more difficult and in some cases a highly uncertain process. This method is likely to be more efficient when the bulk of the data error is in the input, rather than in the output. If the matching can be achieved, the erroneous roots are removed.

Garvey's numerical experimentation indicated that this method of root matching would give an *error in the derived unit hydrograph* of 4.6% (mean of 36 cases) for 10% systematic error in the data and an error of 21.5% (mean of 36 cases) for 10% random error in the data. Table 4.7 shows that in the case of random error the value of the error in the derived unit hydrograph is, as might be expected, more sensitive to the location of the data error than to the shape of input.

Root matching

Table 4.7 shows the method of *root matching* to be particularly efficient when all of the error is in the input data.

The original method of matching root values between events on the same catchment due to De Laine (1970, 1975) was applied by Turner (1982) to data from the River Liffey. He noted that the complex roots of the unit hydrograph when plotted on an Argand diagram displayed a characteristic pattern, which he described as a skewed circle. Similar patterns were obtained for other catchments in Ireland and for synthetic data based on a cascade of equal linear reservoirs (Turner et al., 1989). Subjective pattern recognition was used to separate the roots for effective rainfall and the roots for rapid catchment response. This method of root selection overcame the problem of the sensitivity of the root pattern to such factors as random error in the input data and the effect of the length of the runoff hydrograph, to which the root matching method was susceptible.

Research has since been carried out on a further extension, by separating from the total runoff hydrograph, the three components of rainfall, slow response and quick response.

The iterative approach of Obled and his students (Sempere Torres, Rodriguez, and Obled, 1992) to the removal of "input losses", alternating between solving the detection and identification problems, can be compared with de Laine's method.

4.6 UNIT HYDROGRAPH DERIVATION BY HARMONIC ANALYSIS

Harmonic analysis

As mentioned in Chapter 3, O'Donnell (1960) applied *harmonic analysis* to the problem of unit hydrograph derivation. If the input, pulse response and output of a linear time-invariant system are all represented as finite Fourier series, the k-th complex Fourier coefficients of the output (C), of the input (c), and of the pulse response (γ) are connected by the linkage equation (3.53)

$$C_k = nc_k\gamma_k \tag{4.15}$$

The similarity between equations (4.13) and (4.15) is due to the fact that the coefficients in equation (4.15) for any particular value k correspond to equation (4.13) with a value of z given by

$$z = \exp\left(\frac{2\pi ik}{n}\right) \tag{4.16}$$

where n is the number of data points for the function concerned. The efficiency of this method of identification is distinctly improved if use is made of the *Fast Fourier transform* algorithm (Brigham, 1974). If the trigonometrical form of the Fourier series is used, the linkage equation takes the form of equation (3.54) given in Chapter 3 above.

Fast Fourier transform

Harmonic coefficients

If a full set of harmonic coefficients is used in the analysis, the derived unit hydrograph will contain all of the frequencies in the data up to half the sampling frequency. It will reproduce both the *signal* represented by the underlying true unit hydrograph and the *noise* represented by the data error. If the *series is truncated*, then the expectation will be that, for the heavily damped systems encountered in hydrology, the removal of the high frequency components will remove the greater part of the noise without undue impairment of the underlying signal. This effect is shown clearly in the results of numerical experimentation due to Garvey (1972) and shown in Table 4.8.

It can be seen from Table 4.8, that for the case of error free data, the effect of truncation is to increase the error in the derived unit hydrograph as the error due to roundoff is small. The information lost by truncation is almost entirely loss of information in regard to the true signal. In the case of 10% systematic error, the truncation of the series removes some of the noise as well as of the signal. Over a wide range of length of series there is an *approximate balance* between the increase in error due to loss

Table 4.8. Effect of length of harmonic series.

No. of terms	Mean absolute error in unit hydrograph of % of peak			
	Error-free	Systematic error	Random error	Mean for 10% error
37	0.01	6.7	21.3	14.0
31	0.08	6.7	20.9	13.6
27	0.15	6.7	18.9	12.8
17	0.7	5.9	11.9	8.9
15	1.0	**4.5**	10.6	7.6
13	1.5	**4.5**	9.6	7.1
11	2.3	4.7	8.8	6.8
9	3.4	5.3	7.8	**6.6**
7	5.5	6.9	**7.1**	7.0
5	8.3	9.2	8.8	9.0

Table 4.9. Harmonic analysis for 10% random error.

Type of error	Mean absolute error as % of peak	
	N = 37	N = 9
Input only	3.5	5.1
Output only	34.7	11.5
Equally divided	25.7	6.9
Average for 6 cases	21.3	7.8

of signal information and the decrease in error due to the filtering out of information due to noise from data errors. In the case of 10% random error, the effect of the noise due to data error is still more marked, and the optimum result is obtained for a series as short as seven terms.

Random error

A more detailed examination of the results for *random error* reveal that the errors in the derived unit hydrograph are greater when the data error is in the output rather than in the input. This is shown in Table 4.9 for the complete series of 37 terms and for a truncated series of 9 terms.

It will be noted that for the average of the two cases of error in the input only, the effect of *truncating the series* from 37 terms to 9 terms is to increase the error in the derived unit hydrograph. This may be contrasted with the other two cases in Table 4.9, where truncation sharply reduces the error.

Systematic error

A detailed examination of the six *cases of systematic error* reveals a similar variation of the effect of truncation with type of error. With 10% error in the input, due to rain-gauge lack of synchronisation, reduction of the series length from 37 terms to 9 terms reduced the error in the unit hydrograph from 16.4% to 5.3%. The same reduction in length increases the mean error for the other five cases of systematic error (as listed in section 4.2 above) from 4.7% to 5.6%. Because of the large reduction in the case of lack of *rain-gauge synchronisation*, the overall effect on the mean of all six cases is a reduction from 6.7% to 5.3% as indicated in Table 4.8.

4.7 UNIT HYDROGRAPH DERIVATION BY MEIXNER ANALYSIS

For the heavily damped systems encountered in hydrology, it has been suggested that exponential type functions would be more suitable as a basis for expansions in the transform approach, than trigonometrical functions. In Chapter 3 it was pointed out that Laguerre functions have been used for continuous systems and Meixner functions for discrete systems for this reason. The ordinary *Meixner function* (Dooge, 1966; Dooge and Garvey, 1978) is defined by equation (3.49)

$$f_n(s) = \left(\frac{1}{2}\right)^{(s+n+1)/2} \sum_{k=0}^{n} (-1)^k \binom{n}{k}\binom{s}{k} \tag{4.17}$$

Meixner series

Linkage equation

which are orthonormal over the range $s = 0, 1, 2 \ldots \infty$. If the input, the pulse response and the output functions are all expanded as Meixner series, then the coefficients of the output (A), the coefficients of the input (a) and the coefficients of the pulse response (α) are connected by the *linkage equation* (3.57):

$$A_p = \sum_{k=0}^{p} (\sqrt{2} \cdot a_{p-k} - a_{p-k-1})\alpha_k \tag{4.18}$$

The matrix of coefficients formed from the input coefficients (a) which multiplies the vector of unknown coefficients of the pulse response (α) in equation (4.18) is seen to be of the same form as the matrix defined by equation (4.2) formed from the input vector. Any reliable method of matrix inversion, or of optimisation, can be used to solve equation (4.18).

Meixner analysis is found to be even more sensitive than harmonic analysis to the number of terms in the series, for cases where there is error in the input and the output data. Numerical results for this as obtained by Garvey (1972) are shown in Table 4.10 where the linkage equation is solved by forward substitution of equation (4.18).

It can be seen from the above that, even for error free data, there can be considerable errors in the derived unit hydrograph using a long Meixner series. This difficulty arises partly from problems in the generation of high-order Meixner series (Dooge and Garvey, 1977) and can be overcome to some extent by the use of time-scaling. However, it is also noteworthy that for the error free data the unit hydrograph can be determined at an accuracy of about 1% for series of any length in-between 5 terms and 15 terms. Only when the length of series is reduced to 2 or 3 terms does the

Table 4.10. Effect of length of series on Meixner analysis. (Forward substitution solution of equation 4.18.)

Length of series	Mean absolute error as % of peak			
	Error-free	Systematic error	Random error	Mean for 10% error
25	21.0	19.8	28.5	24.2
23	16.2	16.2	27.0	21.6
20	7.3	13.6	23.2	20.0
15	1.0	10.5	18.3	14.4
10	**0.3**	7.9	13.1	10.5
8	0.7	6.1	10.0	8.1
7	0.8	5.7	8.6	7.2
6	1.0	5.2	7.1	6.2
5	1.2	4.8	6.3	5.6
4	1.7	**4.4**	**5.2**	**4.8**
3	8.3	9.5	9.2	9.4
2	11.7	11.6	12.2	11.9

Table 4.11. Effect of series length on Meixner analysis. (Least squares solution of equation 4.18.)

No. of ordinates in pulse response	Mean absolute error as % of peak			
	Error-free	Systematic error	Random error	Mean for 10% error
25	21.0	19.8	28.5	24.2
23	0.4	12.5	20.3	16.4
20	**0.1**	6.8	17.3	14.1
15	0.2	4.7	12.5	8.6
10	0.4	4.0	7.3	5.7
8	0.7	3.8	6.0	4.9
7	0.7	**3.8**	5.6	4.7
6	1.1	3.8	5.2	4.5
5	1.1	3.7	4.8	**4.3**
4	2.1	4.2	**4.4**	4.3
3	7.4	7.5	8.0	7.8
2	9.6	9.1	9.9	9.5

error become appreciable. With 10% error in the data, the error in the derived unit hydrograph will not exceed roughly 5%, even for a series containing only 4 or 5 terms.

Alternatively, we can

(a) derive twenty-five Meixner coefficients for the input and twenty-five Meixner coefficients for the output; and
(b) solve for less than twenty-five Meixner coefficients of the unit hydrograph by least squares (L.S.).

The results are shown in Table 4.11.

Table 4.12. Meixner analysis for 10% random error.

Type of error	Mean absolute error as % of peak		
	$N = 25$	$N = 5$ (F.S.)	$N = 5$ (L.S.)
Input only	24.5	4.7	2.9
Output only	34.5	8.2	7.0
Equally divided	26.6	5.9	4.6
Average for 6 cases	28.5	6.3	4.8

It can be seen from Table 4.11 that the error in the derived unit hydrograph can be considerably reduced for a long series and somewhat reduced for a shorter series by using *least squares* rather than forward substitution to solve the linkage equation. The use of the least squares solution naturally involves more computer time and the choice between the two methods is not immediately obvious but depends on the circumstances of the individual problem.

As in the case of harmonic analysis the level of error in the unit hydrograph depends on the type of error in the data. Table 4.12 shows the results when the cases of 10% random error are grouped according to whether the error is in the input or the output.

Again the best results are obtained when the error occurs only in the input. For 10% systematic error there is a similar variation between the errors for the six types of error. For the 5-term Meixner solution by forward substitution the error in the unit hydrograph varies from 3.3% for an error in the rating and 3.5% for an error in rainfall synchronisation, to 8.1% for an error due to assuming a uniform loss rate. For the 5-term Meixner solution by least squares, the error in the unit hydrograph varied from 1.6% for an error in rainfall synchronisation and 2.4% for an error in the rating curve to 5.8% for an error due to using a uniform loss rate.

The performance of the various transform methods of unit hydrograph derivation discussed in sections 4.5, 4.6 and 4.7 is summarised in Table 4.13.

It can be seen from Table 4.13 that all the methods, except polynomial division, give comparable and reasonably satisfactory results for the case of 10% systematic error. However, for the case of 10% random error, Meixner analysis is somewhat better than harmonic analysis and both are substantially better than root-matching.

4.8 OVERALL COMPARISON OF IDENTIFICATION METHODS

As mentioned at the beginning of this chapter, one of the main purposes of the chapter is to compare the various approaches to the derivation of the unit hydrograph. We have seen that many methods, which give

Table 4.13. Summary of transform methods.

Method of identification	Mean absolute error as % of peak			
	Error-free	Systematic error	Random error	Mean for 10% error
z-transform (polynomial division)	0.9×10^{-3}	27.0	45.5	36.4
z-transform (root-matching)	0.1	4.6	21.5	13.1
Harmonic analysis ($N = 9$)	3.4	5.3	7.8	6.6
Meixner analysis (F.S.*, $N = 5$)	1.2	4.8	6.3	5.6
Meixner analysis (L.S.**, 25´5)	1.1	3.7	4.8	4.3

*forward substitution. **least squares.

Table 4.14. Overall comparison of identification methods.

Method of identification	Mean absolute error as % of peak			
	Error-free	Systematic error	Random error	Mean for 10% error
Collins method	0.09×10^{-3}	5.8	27.7	16.8
Least squares	0.29×10^{-3}	6.6	21.5	14.1
Regularisation	3.1×10^{-3}	4.3	11.5	7.9
z-transform (roots)	0.12×10^{-3}	4.6	21.5	13.1
Harmonic analysis (9 terms)	3.4×10^{-3}	5.3	7.8	6.6
Meixner analysis (forward substitution, 5)	1.2×10^{-3}	4.8	6.3	5.6
Meixner analysis (L.S., 25/5)	1.1×10^{-3}	3.7	4.8	4.3

Error-free data

excellent results for error-free data, are not robust in the presence of errors in the input and output data and consequently are of no real use in applied hydrology.

The methods that keep the error in the derived unit hydrograph under control, all have included in them, either implicitly or explicitly, some form

Numerical filtering

of numerical filtering. The most useful methods are those, whose filtering action is such, that it removes a large proportion of the noise due to errors in the data, without unduly affecting the underlying signal represented by the "true" unit hydrograph. Table 4.14 based on the numerical experiments by Garvey (1972) shows an overall comparison of the more robust of the methods covered in the preceding sections.

As we have seen in the individual examination of the various approaches, there are several methods which can keep the error in the derived unit hydrograph reasonably small for the case of 10% systematic error. But a number of these are not so robust for the case of 10% random error in the data. Since a method for use in applied hydrology must be capable of standing up to both systematic and random errors, methods capable of dealing efficiently with both types of error are to be preferred.

Table 4.15. Comparison of relative CPU times for different methods.

Method of identification	Percentage error for 10% error		Relative CPU time
	Systematic	Random	
Collins method	5.8	27.7	1
Least squares	6.6	21.5	2.3
Regularisation	4.3	11.5	13.7
z-transform (roots)	4.6	21.5	9.9
Harmonic analysis ($N = 9$)	5.3	7.8	0.5
Meixner analysis (F.S., 5)	4.8	6.3	0.5
Meixner analysis (L.S., 25/5)	3.7	4.8	1.7

Table 4.16. Effect of level of error.

Method of identification	Mean absolute error as % of peak			
	Error-free data	5% error in the data	10% error in the data	15% error in the data
Forward substitution	0.85×10^{-3}	42.4	45.4	50.3
Collins method	0.09×10^{-3}	10.6	27.7	38.6
Least squares	0.29×10^{-3}	7.8	21.5	34.2
Regularisation	3.1×10^{-3}	5.6	11.5	17.8
Harmonic analysis (9 terms)	3.4×10^{-3}	5.1	7.8	14.2
Meixner analysis (L.S., 25/5)	1.1×10^{-3}	3.1	4.8	6.9

In choosing between methods giving approximately the same degree of accuracy we would naturally choose a method which involves less computer time. Accordingly, Table 4.15 presents (a) the relative CPU time for the more promising methods, and (b) the mean absolute error in the derived unit hydrograph as a percentage of peak, for the mean of the six cases of 10% systematic error, and the mean of the six cases of 10% random error in the data.

It is clear from Table 4.15 that the two methods based on orthogonal function transformation are superior to the other methods, both in regard to accuracy of the derived unit hydrograph and economy of computer operation.

Garvey (1972) also experimented on the effect of the level of error in the data on the mean absolute error in the derived unit hydrograph. The computations were carried out for both systematic and random error. Only the results for random error are summarised here, as these are the most significant (Table 4.16).

It can be seen from Table 4.16 that the results for the other levels of error give the same ranking of the methods as for the 10% level of error discussed in detail above.

CHAPTER 5

Linear Conceptual Models of Direct Runoff

5.1 SYNTHETIC UNIT HYDROGRAPHS

In Chapter 4 we discussed the black-box analysis of linear time-invariant systems with particular reference to the component of the system response concerned with the direct storm response to precipitation. This particular component is illustrated in Figure 2.3 (Simplified catchment model) as the component of the total catchment response which converts effective precipitation P_e to storm runoff Q_s. It was mentioned in Section 1.1 and again in relation to Figure 2.4 (Models of hydrological processes) that the black-box approach was one of the three general approaches to the prediction of the system output. In the present chapter we are concerned with the approach to the prediction of direct storm response based on simple *conceptual models* i.e. with an approach which is intermediate between the black-box approach discussed in Chapter 4 and the approach based on the equations of mathematical physics as applied to hydrological phenomena.

Conceptual models

It was noted in Chapters 1 and 2 that the term "conceptual model" can be used broadly to cover both black-box models based on systems analysis and mathematical models based on continuum mechanics. In this book, however, the term conceptual model will be used in the restricted sense of models that are formulated on the basis of a simple arrangement of a relatively small number of elements, each of which is itself a simple representation of a physical relationship.

The most widely used conceptual elements of the direct storm runoff component are *linear channels* and *linear reservoirs*. These elements represent a separation and a concentration of the two distinct processes of *translation* and *attenuation*, which are combined together in the case of unsteady flow over a surface or in an open channel (Dooge, 1959). The conceptual model of a *cascade of equal linear reservoirs*, each with a *lateral inflow*, corresponds to the assumption that the system function of black-box analysis may be approximated by the ratio of two polynomials (i.e. by a *rational function*). Such a conceptual model is, therefore, closely related to the black-box approach. On the other hand, the conceptual model based on the convective-diffusion analogy corresponds to a linearised version

Rational function

of the St. Venant equations for the case of a vanishingly small Froude number. It may be said to represent a transition between a conceptual model of the direct storm response as a lumped system and a simplified distributed mathematical model based on the equations for unsteady free-surface flow.

Conceptual models of direct storm response first emerged in hydrology in connection with the problem of *synthetic unit hydrographs*. It is a commonplace of applied hydrology that the important problems seem to arise in relation to catchment areas for which little or no information is available. Accordingly, it is not sufficient to be able to derive a unit hydrograph for a catchment area, in which there are records of rainfall and runoff, using one of the methods discussed in Chapter 4. It is also desirable to develop procedures for the derivation of synthetic unit hydrographs and hence the prediction of storm runoff for *ungauged catchments*. This can be attempted, if rainfall and runoff data are available for similar catchments in the same general region.

Ungauged catchments

The basic approach used is as follows:

(1) derive the unit hydrographs for the catchments in the region for which records are available;
(2) find a correlation between some defined parameters of these unit hydrographs and the catchment characteristics;
(3) use this correlation to predict the parameters of the unit hydrograph for catchments, which have no records of streamflow, but for which the catchment characteristics can be derived from topographical maps.

Sherman (1932a) published the basic paper on the unit hydrograph approach in 1932. In the same year he published a second paper, which was concerned with the relationship between the parameters of the unit hydrograph and the two important catchment characteristics of area and slope (Sherman, 1932b). However, it may be said that the first synthetic unit hydrographs were in fact derived more than ten years before the concept of the unit hydrograph itself appeared in print.

Rational method

In 1921 Hawken and Ross modified the classical *rational method* (Mulvany, 1850) to include the effect of non-uniform rainfall distribution by the use of *time–area–concentration curves*. These curves were estimated on the basis of the time of travel from various parts of the catchment to the outlet, as computed by hydraulic equations for steady flow. The time–area–concentration curve and the design storm were plotted to the same time scale but with time running in the opposite direction in the two curves. The curve for the design storm and the curve for the time–area–concentration curve were then superimposed, the products of corresponding ordinates taken and summed together to obtain the runoff at any given time. This was in effect a graphical method of carrying out the four operations of convolution: shift, fold, multiply, integrate, described in Section 3 of Chapter 1.

The time–area–concentration curve in such a procedure, performed the same function as the instantaneous unit hydrograph in the linear systems approach. Thus the time–area–concentration curve, whose shape was computed on the basis of catchment characteristics, was in fact a synthetic unit hydrograph. In each case, the time–area–concentration curve was built up by application of hydraulic equations to the information available for the particular catchment. Hence, each of these unit hydrographs was unique to the catchment concerned.

The time–area–concentration curve, as used in the rational method, was based solely on the estimate of the time of translation over the ground and in channels. The results obtained using the method tended to overestimate the peak rate of discharge from the catchment due to neglect of runoff attenuation by surface storage, soil storage and channel storage. Following the introduction of unit hydrograph methods, numerous attempts were made to allow for the storage affect in one way or another.

Clark (1945) suggested that the instantaneous unit hydrograph could be derived by routing the time–area–concentration through a single element of linear storage. In this case also, each unit hydrograph would be unique, but the variation between them would be reduced, and the difference in catchment characteristics smoothed out, depending on the degree of damping introduced by the storage routing. O'Kelly, Nash and Farrell, working in the Irish Office of Public Works, found that there was no essential loss in accuracy in the synthesis of unit hydrographs, if the routed time–area–concentration curve was replaced by a *routed isosceles triangle.*

In the case of a routed triangle, the unit hydrographs are no longer unique, but belong to a family of two-parameter curves. The parameters required to characterise such a unit hydrograph are the base of the isosceles triangle T and the *storage delay time K* of the *linear storage element.* Thus the line of development, which started out by treating each unit hydrograph as unique, had in time developed into an empirical procedure in which there were only two unknown parameters.

In contrast to the time–area methods, which treated each catchment as unique, a number of hydrologists, during the 1930s and later, suggested that a unique shape could be used to represent the unit hydrograph. This would take the form of a dimensionless unit hydrograph and only one parameter would be required in order to determine the scale of an actual unit hydrograph. Since the volume under the unit hydrograph was normalised to unity, a change in the time scale was automatically compensated for by a corresponding change in the discharge scale.

Universal shapes for the unit hydrograph

A number of such *universal shapes for the unit hydrograph* were proposed in the literature (e.g. Commons, 1942). As further studies were made of the synthetic unit hydrographs, it was realised that the one-parameter method was not sufficiently flexible, and that at least two parameters were required for adequate representation of derived unit hydrographs. Such a two-parameter representation would require the use of a family of curves from which the unit hydrograph shape might be chosen. Since it is easier to represent the two-parameter model by an equation, rather than by a family of curves, the natural development of this approach was towards the suggestion of empirical equations, which would represent all unit hydrographs.

It is remarkable that people working in a number of different countries turned to the same empirical equation for the representation of the unit hydrograph. The equation in question was the two-parameter *gamma distribution* or Pearson type III empirical statistical distribution. This was suggested by Edson in 1951 but the reasoning on which he based his proposal was faulty. In 1956, Sato and Mikawa suggested the one-parameter conceptual model corresponding to two linear reservoirs in series, and in the following year, Sugawara and Maruyama suggested a three-parameter model consisting of three cascades of equal linear reservoirs in parallel, containing one, two and three reservoirs respectively.

Two-parameter gamma distribution

Nash (1958) suggested the two-parameter gamma distribution as having the general shape required for the instantaneous unit hydrograph and pointed out that the gamma distribution could be considered as the impulse response for a cascade of equal linear reservoirs. He suggested that the number of reservoirs could be taken as non-integral if required. Thus the line of development in synthetic unit hydrographs, which started from the point of view of each unit hydrograph being the same, also ended in the proposal of a two-parameter shape for the unit hydrograph.

The two lines of development in regard to synthetic unit hydrographs are shown in Figure 5.1. It can be seen that the approach, which started with time–area assumption that every unit hydrograph was unique, ended with the routing of a fixed triangular shape through a single linear reservoir.

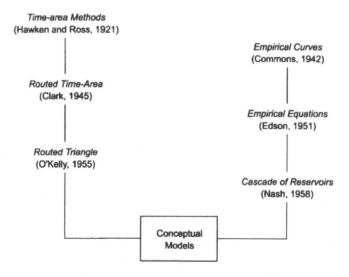

Figure 5.1. Development of synthetic unit hydrographs.

Similarly the line of development, which started with the assumption that there was a single universal shape for the unit hydrograph, led to the representation of the unit hydrograph by a cascade of equal linear reservoirs.

It is clear that both the method of routing a triangular inflow through a single linear reservoir to obtain the instantaneous unit hydrograph, and the use of a cascade of equal linear reservoirs to simulate the instantaneous unit hydrograph, are both conceptual models in the sense defined in Chapter 1. Thus the two different approaches to synthetic unit hydrographs (one based on each unit hydrograph being unique and the other based on there being a universal shape for the unit hydrograph) both emerged under the pressure of fitting the hydrological facts to the proposal of a two-parameter conceptual model.

From this time on, the way was open to represent the unit hydrograph by a wide variety of conceptual models.

There is no limit to the number of conceptual models that can be devised. Indeed, a grave defect in hydrological research in recent years has been the proliferation of conceptual models, without a corresponding effort to devise methods of objectively comparing models, and developing criteria for the best choice of model in a given situation. A conceptual model does not become a synthetic unit hydrograph in the full sense, until its parameters are correlated with catchment characteristics. This topic is outside the scope of the present book, but has been dealt with elsewhere by Dooge (1973), pp. 197–206.

5.2 COMPARISON OF CONCEPTUAL MODELS

If a conceptual model is to be used to represent the action of a catchment area on effective rainfall, it is necessary to choose

(1) a conceptual model; and
(2) values of the parameters for the chosen model.

If the model is chosen at random, and if the parameters are chosen by trial and error on the basis of fitting the runoff, the approach may be even more subjective than the early method of deriving a unit hydrograph by trial and error. It was seen in Chapter 4, and will be seen again in relation to conceptual models in the present chapter, that the matching of the output is no guarantee that the derived unit hydrograph resembles closely the actual unit hydrograph. Accordingly it is necessary to develop, if possible, *objective methods* for choosing a conceptual model and for determining the parameters of that model.

Objective methods

The first question to be considered is the manner in which the unit hydrograph may be described. It may of course be described in terms of the derived ordinates at some specified interval as in the black-box approach.

In this case, the parameters to be determined are the interval used and the values of a sufficient number of ordinates to describe adequately the shape of the unit hydrograph. In order to develop synthetic unit hydrographs, it would be necessary to correlate all of these ordinates with catchment characteristics. Since this is likely to prove difficult, synthetic methods in the past have attempted to correlate other characteristics, such as the peak of the unit hydrograph or some measure of the time of occurrence of the peak.

Describing a unit hydrograph

The problem of *describing a unit hydrograph* in a compact fashion is the same problem as that of describing a frequency distribution in statistics. Nash (1959) suggested that the *moments of the instantaneous unit hydrograph* should be used to

(1) describe its shape; and
(2) compare various derived unit hydrographs, or conceptual models.

Since the moments of the unit hydrograph can be determined from the moments of the effective precipitation and the direct storm runoff (as also pointed out by Nash), the moments of the instantaneous unit hydrographs can be determined from the available data without deriving the full unit hydrograph. The linkage equation for cumulants and moments (3.75) shows that the first three moments of the unknown unit hydrograph can be found by subtracting the first three moments of the input from the first three moments of the output. Nash also suggested the use of *dimensionless moments* for representing the shape of a unit hydrograph, which is free from the effect of scale.

The moments used in systems hydrology are the moments of the various functions with respect to time. The *moments about the time origin* of a function $f(t)$ are defined as expression (3.69) in Chapter 3

$$U'_R(f) = \int_0^\infty f(t) t^R \, dt \tag{5.1}$$

and the moments about the centre of area are defined as

$$U_R(f) = \int_0^\infty f(t)(t - U'_1)^R \, dt \tag{5.2}$$

The relationship between the moments about the origin defined by equation (5.1) and the moments about the centre defined by equation (5.2) can be found by expanding the term $(t - U'_1)^R$ in equation (5.2). As will be seen later, the moments can be used for determining the parameters of conceptual models, as well as providing the basis for the comparison of the models.

Dimensionless moments or shape factors may be defined by

$$s_R = \frac{U_R}{(U'_1)^R} \tag{5.3}$$

where U_R is the R-th moment about the centre and U'_1 is the first moment about the origin. The shape factors defined by equation (5.3) are dimensionless. If the area under the unit hydrograph is also normalised to unity, they will characterise only the shape of the unit hydrograph.

Shape-factor diagram

A diagram, on which the dimensionless third moment s_3 is plotted against the dimensionless second moment s_2, may be referred to as a *shape-factor diagram*. Other dimensionless moments could be plotted against one another, but the most useful results are found by plotting in terms of the two lowest dimensionless moments. Conceptual models with one, two or three parameters can be represented on a shape-factor diagram. A one-parameter model can be represented by a point, a two-parameter model by a line, and a three-parameter model by a region. Results for examples of all three types are discussed in Dooge (1977, pp. 92–98), which deals with ten numbered one-parameter models (1–10), ten two-parameter models (11–20), and four three-parameter models (21–24).

The use of a *shape-factor diagram* to compare different conceptual models is illustrated by Figure 5.2. This figure shows the plotting of the dimensionless third moment s_3 for the two conceptual models mentioned in the discussion of Figure 5.1, namely, the *routed isosceles triangle* (model 14) and the *cascade of equal linear reservoirs with upstream inflow* (model 16). The conceptual model based on the convective-diffusion analogy (model 20), mentioned earlier in this Chapter, is also plotted in Figure 5.2. These three dissimilar conceptual models plot quite close to one another on a shape factor diagram, and therefore are likely to be quite similar in their ability to match the shape of a derived unit hydrograph.

The *shape-factor diagram* can also be used for the *plotting* of derived unit hydrographs. Once the moments are known for the unit hydrograph of a particular storm on a particular catchment area, this unit hydrograph can

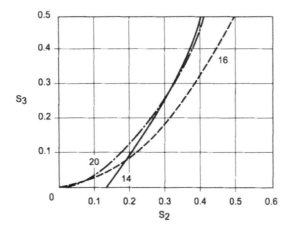

Figure 5.2. Comparison of conceptual models with 2 parameters.

be plotted as a point on a shape-factor diagram. If data are available from a number of catchments in a region, they can be plotted on a shape-factor diagram, and the results used to judge the ability of various conceptual models to represent unit hydrographs for that region.

If all the plotted points for derived unit hydrographs are clustered around a single point in the diagram, a one-parameter model plotted at this point, would be sufficient to represent all the unit hydrographs in the region. If the points plotted along a line, one could represent these unit hydrographs by a two-parameter conceptual model, whose characteristic line on a shape factor diagram passed close to all the points. If the plotted points for the derived unit hydrograph filled a compact region on the shape-factor diagram, only a three-parameter conceptual model spanning that region on the diagram would be capable of simulating adequately all these derived unit hydrographs.

5.3 CASCADES OF LINEAR RESERVOIRS

It might be thought that there would be no difficulty in fitting any set of regional hydrographs, and all that would be required, would be to increase the number of parameters until a fit is obtained. However, *both analytical studies*, which will be summarised here, *and the results of numerical experimentation, indicate that this is not so.*

System function   It was shown in Chapter 3 (see expression (3.79)) that if the *system function* (i.e. the Laplace transform of the impulse response) can be represented in a quite general way by the *ratio of two polynomials*, i.e. by a *Padé approximation* of $H(s)$ (Ralston and Wilf, 1960, p. 13) we have

$$Y(s) = H(s)X(s) = \frac{P_m(s)}{Q_n(s)}X(s) \tag{5.4}$$

which represents the relationship between input and output in the transform domain. Since P_m and Q_n are polynomials, we can use the fundamental theorem of algebra to express them as products of m factors and n factors respectively. Thus we can write

$$H(s) = \frac{\prod_{i=1}^{m}(s - r_i)}{\prod_{j=1}^{n}(s - r_j)} \tag{5.5}$$

where r_i represents a root of the polynomial P_m and r_j represents a root of the polynomial Q_n. Though the above equation has been derived completely on the basis of the black-box approach with the assumption that the system function is a rational function, it can be interpreted in terms of a conceptual model based on a *cascade of linear reservoirs*.

This discussion is presented in two parts. In the first part, we consider the special case where P_m is a constant and we show that the corresponding cascade has an inflow into the first reservoir only of the cascade (upstream

Upstream inflow

inflow). In the second part, we consider the more general case which is stable ($m \leq n$). This corresponds to a cascade with an inflow into each reservoir of the cascade (lateral inflow).

If the system function takes the special form where only the denominator contains powers of s, we have as a special case of equation (5.5)

$$H(s) = \frac{\prod_{j=1}^{n} (-r_j)}{\prod_{j=1}^{n} (s - r_j)} \tag{5.6}$$

When the system satisfies a conservation law, the area under the impulse response function $h(t)$ must be unity (see the discussion following expression (1.8) and the proof in Chapter 3 following expression (3.67)). The Laplace Transform (3.61) of $h(t)$ satisfies this constraint, if and only if, $H(0) = 1$. The numerator of equation (5.6) must take the indicated form, so that $H(0) = 1$. Hence, equation (5.6) can be simplified to

$$H(s) = \prod_{j=1}^{n} \left[\frac{1}{1 - s/r_j} \right] \tag{5.7}$$

It is clear that the system function represented by equation (5.7) is the product of a series of factors of the form

$$H_j(s) = \left[\frac{1}{1 - s/r_j} \right] \tag{5.8}$$

Consequently, the impulse response in the time domain can be obtained by convoluting the individual impulse responses corresponding to the system function represented by equation (5.8). Inverting equation (5.8) to the time domain we obtain

$$h_j(t) = -r_j \exp(r_j t) U(t) \tag{5.9}$$

where $U(t)$ is the unit step function. For heavily damped systems such as occur in hydrology, the values of the roots r_j must be real rather than complex, as oscillations would otherwise occur in the response function. Equally, the values of r_j must be negative, since otherwise the impulse response would grow without limit, indicating an unstable system.

Single linear reservoir

It is now necessary to show that equation (5.9) represents the response of *a single linear reservoir* i.e. of a reservoir for which we have the relationship between storage and outflow, given by

$$S(t) = KQ(t) \tag{5.10}$$

If we write the equation of continuity as

$$I(t) - Q(t) = \frac{d}{dt}[S(t)] \tag{5.11}$$

we can incorporate the storage relationship from equation (5.10) and write

$$I(t) - Q(t) = K\frac{\mathrm{d}}{\mathrm{d}t}[Q(t)] \tag{5.12}$$

which can be written as

$$(1 + KD)Q(t) = I(t) \tag{5.13}$$

where D is the differential operator.

On transforming this equation to the Laplace transform domain, we obtain

$$Q(s) = \left(\frac{1}{1 + Ks}\right)I(s) = H(s)\,I(s) \tag{5.14}$$

so that the system function is seen to be of the same form as equation (5.8) above. Accordingly, we can interpret equation (5.8) as representing *a single linear reservoir with a storage delay time K equal to minus the reciprocal of r_j.*

Cascade of two reservoirs

We now take a *cascade of two reservoirs* i.e. two reservoirs in series, in which the output from the first, becomes the inflow to the second. We can readily determine the response function for this system i.e. we seek the output from the second reservoir for a delta function input to the first. The output from the second reservoir will be the convolution of the impulse response of the second reservoir with the input to the second reservoir. That input is equal to the output from the first reservoir, which is the impulse response of the first reservoir. Hence, the *impulse response* for the total system must be the convolution of the separate impulse responses of the two reservoirs which are in series. But convolution in the time domain is transformed to multiplication in the Laplace transform domain; accordingly, the system function or Laplace transform of the system response for the two reservoirs in series will be given by

$$H(s) = \left[\frac{1}{1 + K_1 s}\right]\left[\frac{1}{1 + K_2 s}\right] \tag{5.15}$$

Applying this reasoning to a cascade of n reservoirs, we realise that the system function represented by equation (5.7) corresponds to a cascade of reservoirs whose storage delay times are equal to the reciprocals of the roots of the polynomial Q_n. Thus for a *cascade of unequal reservoirs* the system function will be given by

Cascade of unequal reservoirs

$$H(s) = \frac{1}{\prod_{j=1}^{n}(1 + K_j s)} \tag{5.16}$$

Clearly, the order in which the reservoirs are arranged in the cascade is of no consequence with respect to its final output. Hence there are $n!$ equivalent cascades — the number of permutations of the n-reservoirs — which produce identical outputs for all possible upstream inflows. The result is true for all linear time-invariant systems, which consist of sub-systems in series with upstream input, since convolution is commutative, and associative (Section 1.3).

It can be shown (Dooge 1959) that the corresponding impulse response in the time domain is given by

$$h(t) = \sum_{j=1}^{n} \frac{(K_j)^{n-2} \exp(-t/K_j)}{\prod_{i=1}^{n-1}(K_i - K_j)}, \quad i \neq j \quad \text{and} \quad K_i \neq K_j \tag{5.17}$$

so that the *impulse response in the time domain* consists of a sum of exponential terms. Note that all storage delay times must be different, and that i may not equal j in forming the product of their differences in the denominator. The case of "equal delay times" requires a different inversion of the system function (5.16).

n equal linear reservoirs

In the particular case of *n equal linear reservoirs* the system function given by equation (5.16) will take the special form

$$H(s) = \frac{1}{(1 + K s)^n} \tag{5.18}$$

and it can be shown that the impulse response in the time domain is given by

$$h(t) = \frac{1}{K} \frac{\exp(-t/K)(t/K)^{n-1}}{(n-1)!} \tag{5.19}$$

which is the *gamma distribution*.

It might be thought that the conceptual model represented by equation (5.19), which contains only two parameters, (n the number of reservoirs and K the storage delay time of each) would be markedly inferior in simulation ability to the system represented by equation (5.17), which has n parameters corresponding to the n storage delay times of the unequal reservoirs in the cascade, and for which n can be made as large as we wish. However, it will be shown both analytically and by plotting them on a shape factor diagram, that there is little difference between the two conceptual models.

Since we know the Laplace transform of the impulse response for a *single linear reservoir*, we can use its moment-generating property (see the remark following equation (3.61) and expression (3.72) for the case of the Fourier Transform) to show that the R-th moment of the impulse response about the origin is given by

$$U_R' = (R)! K^R \tag{5.20}$$

or to show that the R-th cumulant is given by

$$k_R = (R - 1)! K^R \tag{5.21}$$

We can apply the theorem of cumulants (3.75) to derive the expression for the *R-th cumulant of a cascade of n reservoirs* as

$$k_R = (R-1)! \sum_{j=1}^{n} (K_j)^R \tag{5.22}$$

R-th shape factor

Accordingly the *R-th shape factor* or *dimensionless cumulant* is given by

$$s_R = \frac{(R-1)! \sum_{j=1}^{n} (K_j)^R}{(\sum_{j=1}^{n} K_j)^R} \tag{5.23}$$

and can be calculated readily when the *n* storage delay times K_j are known. In particular, for the case of *n* equal linear reservoirs we have

$$s_R = \frac{(R-1)! \, nK^R}{(nK)^R} = \frac{(R-1)!}{(n)^{R-1}} \tag{5.24}$$

which does not depend on K. This gives for the dimensionless second moment or second cumulant

$$s_2 = \frac{1}{n} \tag{5.25}$$

and for the dimensionless third moment or third cumulant

$$s_3 = \frac{2}{n^2} \tag{5.26}$$

and for the dimensionless fourth cumulant (which is not equal to the dimensionless fourth moment)

$$s_4 = \frac{6}{n^3} \tag{5.27}$$

A comparison of equations (5.24) and (5.25) indicates that the two-parameter conceptual model of a cascade of equal linear reservoirs whose impulse response is given by equation (5.19) is represented on a (s_3, s_2) shape-factor diagram by the line whose equation is

$$s_3 = 2(s_2)^2 \tag{5.28}$$

This line appears in Figure 5.3 as model 16. Expressions (5.25) and (5.26) are the parametric equations of this line. For the case of a single reservoir, $n = 1$, and $(s_3 = 2, s_2 = 1)$. The origin corresponds to infinitely many reservoirs in the cascade.

The combination of the use of conceptual models based on linear reservoirs (Nash, 1958; Dooge, 1959) with the quantitative laws of catchment *Geomorphological* morphology (Horton, 1945) led to the concept of the geomorphological *unit hydrograph* unit hydrograph, originally based on a network of stochastic elements (Rodriguez-Iturbe and Valdez, 1979), and later reformulated for a network of deterministic elements (Chuta and Dooge, 1990). In the latter study, the basic assumptions made by Rodriguez-Iturbe and Valdez are used to

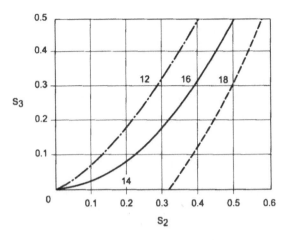

Figure 5.3. Limiting forms of the unimodal cascade. (Model classification number is defined in Chapter 6 and in Appendix A)

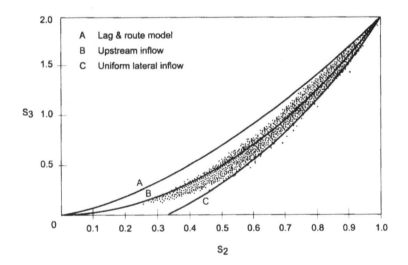

Figure 5.4. Simulation of GUH (Geomorphic Unit Hydrograph).

construct a network of linear reservoirs and to derive explicit equations for the unit hydrograph and for its first three cumulants in terms of the basic parameters, the bifucation ratio R_B, the length ratio R_L, and the drainage basic ratio R_A. A series of Monte Carlo experiments were made by taking random samples in the intervals $2.5 < R_B < 5.0, 1.5 < R_L < 4.1$, and $3.0 < R_A < 6.0$. 11,000 separate combinations were drawn.

This work was extended and generalised by Shamseldin and Nash (1998, 1999). Figure 5.4 shows a plot of the dimensionless third moment against the dimensionless second moment for a series of Monte Carlo experiments for catchments of order 2, 3, 4, 5 and 6 (Shamseldin and Nash, 1999, p. 312). It is seen that the points cluster closely around the line for a cascade of equal linear reservoirs with upstream inflow. A statistical correlation gives the result as

$$s_3 = 2.07s_2^{2.2} \tag{5.29}$$

with a coefficient of determination of 0.983. The deviation between equation (5.29) and equation (5.28) for the Nash cascade is less than 4 per cent at $s_2 = 1$ and decreases to zero for $s_2 = 0$.

5.4 LIMITING FORMS OF CASCADE MODELS

Hölder's inequality

It is possible to show, by using Hölder's inequality, that for a fixed value of s_2, no general cascade of n reservoirs with upstream inflow can have a lower value of s_3 than that given by equation (5.28). Using Jensen's inequality it can also be shown that for a fixed value of s_2 no cascade of n linear reservoirs can have a value of s_3 greater than

$$s_3 = 2(s_2)^{3/2} \tag{5.30}$$

Linear channel followed by a linear reservoir

It can be verified that the latter case corresponds to a conceptual model consisting of a *linear channel followed by a linear reservoir*. The first moment for such a model with a channel delay time T and a storage delay time K is given by

$$U_1' = T + K \tag{5.31}$$

the second moment about the centre by

$$U_2 = K^2 \tag{5.32}$$

and the third moment about the centre by

$$U_3 = 2K^2 \tag{5.33}$$

The dimensionless second moment is given by

$$s_2 = \frac{K^2}{(T+K)^2} = \frac{1}{(1+T/K)^2} \tag{5.34}$$

and the dimensionless third moment by

$$s_3 = \frac{2K^3}{(T+K)^3} = \frac{2}{(1+T/K)^3} \tag{5.35}$$

Eliminating $(1 + T/K)$ from equations (5.34) and (5.35) yields equation (5.30). Consequently, the two-parameter conceptual model of a *linear channel of delay time T* and a *linear reservoir of storage delay time K*, can be represented on a (s_3, s_2) shape factor diagram by equation (5.30).

The two limiting forms for a cascade of linear reservoirs are shown on the shape factor diagram in Figure 5.3, where curve 16 represents the *cascade of equal linear reservoirs*, and curve 12 represents the *cascade of a linear channel and a linear reservoir*; and in Figure 5.4 as curve A and curve C respectively.

It might be thought strange that a conceptual model consisting of a linear channel and a linear reservoir could be considered as a limiting case of a cascade of reservoirs. However, if we consider a cascade of reservoirs in which the number of reservoirs n becomes very large and the storage delay time of each K becomes very small in such a way that the product nK remains finite i.e.

$$nK = T \tag{5.36}$$

where T is the total delay time or lag of the cascade, the system function represented by equation (5.18) takes the form

$$H(s) = \frac{1}{(1 + sT/n)^n} \tag{5.37}$$

The limit of $H(s)$ as n tends to infinity is given by

$$H(s) = \exp(-sT) \tag{5.38}$$

When equation (5.38) is inverted to the time domain we obtain the impulse response

$$h(t) = \delta(t - T) \tag{5.39}$$

which represents a *pure translation* with a time delay of T, i.e. a *linear channel*.

In summary, the limiting cases of the (s_3, s_2) relationship for a *cascade with upstream inflow*, given by equations (5.28) and (5.30) above, define an upper and lower bound on s_3 as follows

$$2(s_2)^{3/2} \geq s_3 \geq 2(s_2)^2 \tag{5.40}$$

The bounds are defined on the interval $0 < s_2 < 1$ and coalesce at the points $(0, 0)$ and $(s_2 = 1, s_3 = 2)$ forming a loop of possible (s_2, s_3) pairs. On the right-hand side of equation (5.40), the point $(s_2 = 1, s_3 = 2)$ corresponds to a single linear reservoir in the cascade; on the left-hand side, the point $(s_2 = 1, s_3 = 2)$ corresponds to $T/K = 0$ in the case of the linear channel (T) and reservoir (K). This result can be generalised for any order of shape factor s_R. If s_R is defined as

$$s_R = k_R/(k_1)^R \tag{5.41}$$

Jensen's inequality

it can be shown, by the use of Jensen's inequality and Hölder's inequality respectively, that

$$(R - 1)! \, (s_2)^{R/2} \geq s_R \geq (R - 1)! \, (s_2)^{R-1} \tag{5.42}$$

for any value of R. Expression (5.40) is the special case of equation (5.42) when $R = 3$.

There still remains the question of whether there is a simple conceptual model that corresponds to the general system function when it is the ratio of two polynomials given by equation (5.5) above. It can be shown that the existence of a polynomial P_m in the numerator in equation (5.4) corresponds to lateral inflows to other reservoirs in the cascade other than the upstream reservoir. This can be illustrated for the case where the numerator polynomial is of the first order and the system function can be written as

$$H(s) = \frac{(1 - s/r)}{\prod_{j=1}^{n} (1 - s/r_j)} \tag{5.43}$$

It is easily verified that this is equivalent to the expansion in partial fractions

$$H(s) = \frac{1 - r_1/r}{\prod_{j=1}^{n} (1 - s/r_j)} + \frac{r_1/r}{\prod_{j=2}^{n} (1 - s/r_j)} \tag{5.44}$$

Clearly the first term on the right hand side of equation (5.44) represents the routing of $(1 - r_1/r)$ times the original inflow through all n reservoirs, and the second term the routing of the remainder of the inflow through the remaining $(n - 1)$ reservoirs. This is equivalent to the inflow being divided between the first and second reservoirs in the cascade in the ratio $(1 - r_1/r)/(r_1/r)$.

Since the system is linear and convolution is commutative and associative, the reservoirs may be arranged in any order without affecting the system response. Accordingly, equation (5.43) corresponds to a cascade with positive lateral inflows as long as one of the values of r_j in the denominator is less than the value of r in the numerator. For higher polynomials in the numerator, further lateral inflow will be generated, the number of lateral inflows (additional to that of the most upstream reservoir) corresponding to the order of the polynomial.

Conceptual models based on cascades with lateral inflows can plot in positions outside the loop in Figure 5.3 formed by curve 16, which corresponds to the cascade of equal linear reservoirs, and curve 12, which corresponds to the case of a linear channel followed by a linear reservoir. Intuition suggests and numerical experimentation seems to confirm that for a given cascade, the two limiting cases for cascades with lateral inflow would be the concentration of all the flow in the upstream reservoir, and on the other hand, an equal distribution of lateral inflow among all the reservoirs. The case of uniform lateral inflow to a cascade of equal linear reservoirs is shown as curve 18 in Figure 5.3. For many years no case of distributed lateral inflow was found which plotted below the curve for uniform lateral inflow (curve 18).

For the case of uniform lateral inflow into a cascade of equal linear reservoirs, the first three cumulants are given by

$$k_1 = U_1' = \frac{(n+1)}{2}K \tag{5.45}$$

$$k_2 = U_2 = \frac{(n+1)(n+5)}{12}K^2 \tag{5.46}$$

$$k_3 = U_3 = \frac{(n+1)(n+3)}{4}K^3 \tag{5.47}$$

so that the values for the shape factors are

$$s_2 = \frac{1}{3}\frac{(n+5)}{(n+1)} \tag{5.48}$$

$$s_3 = 2\frac{(n+3)}{(n+1)^2} \tag{5.49}$$

The notion of the case of uniform lateral inflow as a lower limit for values of s_3 was widely accepted, but no mathematical proof was found. Recently, a whole group of lateral distributions have been found that plot below and to the right of curve 18 in Figure 5.3. The relationship between s_3 and s_2 for uniform lateral inflow is given

$$s_3 = \frac{9s_2^2 - 1}{4} \tag{5.50}$$

where s_2 and s_3 are the dimensionless shape factors defined by equation (5.3) in Section 5.2 above. Using the latter equation to convert equation (5.47) from shape factors to moments we obtain

$$(U_3)(U_1') = \frac{9(U_2)^2 - (U_1')^4}{4} \tag{5.51}$$

In the case of cascades with lateral inflow, it is more convenient to operate in terms of the moments about the origin defined by equation (5.1) in Section 5.1 above. Using equation (5.2) in the same section to make the transformation, we obtain

$$(U_3')(U_1') = \frac{9(U_2')^2 - 6(U_2')(U_1')^2}{4} \tag{5.52}$$

The hypothesis that curve 18 in Figure 5.3 is a universal lower limit therefore becomes

$$4(U_3')(U_1') - 9(U_2')^2 + 6(U_2')(U_1')^2 \geq 0 \tag{5.53}$$

and the problem is either to prove this by using a well-established inequality theorem, as in the case of curves 12 and 16 in Figure 5.3, or to negate it by proof or counterexample.

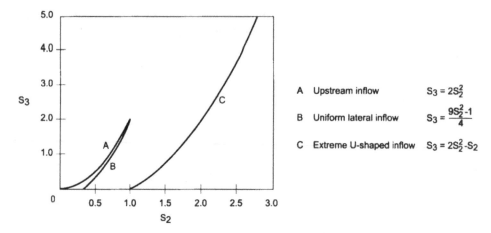

Figure 5.5. Limiting form for U-shaped inflow.

A   Upstream inflow $S_3 = 2S_2^2$

B   Uniform lateral inflow $S_3 = \dfrac{9S_2^2 - 1}{4}$

C   Extreme U-shaped inflow $S_3 = 2S_2^2 - S_2$

A simple counter-example is now available. If we follow the Polya (1945) approach, we tackle first the simplest case of a cascade of two equal linear reservoirs with unequal lateral inflow.

If the inflow to the upstream reservoir is taken as (α), the inflow to the downstream reservoir is given by ($1 - \alpha$). The first three moments about the origin are

$$U_1' = (\alpha + 1)K \tag{5.54}$$

$$U_2' = 2(2\alpha + 1)K^2 \tag{5.55}$$

$$U_3' = 6(3\alpha + 1)K^3 \tag{5.56}$$

Substitution of these values into equation (5.53) reduces the hypothesis of that equation to

$$12(\alpha)^2(2\alpha - 1) \geq 0 \tag{5.57}$$

which is clearly not true for the range $0 \leq \alpha \leq \frac{1}{2}$.

U-shaped distribution

Extension of the analysis to longer cascades reveals that the most serious failure of the hypothesis, occurs when the distribution of lateral inflow along the cascade approaches an extreme U-shaped distribution, corresponding to a long cascade with inflow into the first and last reservoirs only. For such an extreme case the plotting position on the (s_2, s_3) diagram can occur well below and to the right of the case of uniform later inflow shown as curve 18 on Figure 5.3. The later case of uniform inflow is shown as curve B on Figure 5.5 and the case of extreme U-shape, i.e. flow into the upstream and downstream reservoirs only, in an infinite cascade, is shown as curve C. For any length of cascade, n, greater than 2, the $s_3 - s_2$ relationship for an extreme U-shaped distribution follows an unexpected path. The curve starts at the point $s_2 = 1$ and $s_3 = 2$ for the limiting case of vanishingly small input to the upstream reservoir. As the proportion of the inflow into the upstream reservoir increases, the value of s_3 rises above 2, and the value s_2 of increases above 1. Further along the trajectory s_3 peaks and then declines. The curve continues to sweep to the right and after a maximum value of s_2 sweeps back to end on the curve of upstream inflow at $s_2 = 1/n$, $s_3 = 2/n^2$. The extreme case for $n = \infty$ plots on the origin.

In most natural catchments, the width function reflecting the shape of the catchment will be unimodal, rather then U-shaped, and curve 18 will act as a limit in such cases. Hydrologically, such a U-shaped distribution can be avoided by dividing the anomalous catchment and treating the upstream and downstream ends separately. In the limit, we would have two linear reservoirs connected by a linear channel, which would be a reasonable representation of an upper and lower catchment, connected by a steep narrow ravine containing ultra-rapid flow.

CHAPTER 6

Fitting the Model to the Data

The main lesson to be learned from the discussion of Chapter 5 is that there may appear little difference in shape between a well chosen two-parameter conceptual model and one with a larger number of parameters. This would encourage us to attempt to fit unit hydrographs with conceptual models based on two or three parameters, rather than on more complex conceptual models with a large number of parameters.

Principle of parsimony

An additional advantage of using a small number of parameters is that this enables us to concentrate the information content of the data into this small number of parameters, which increases the chances of a reliable correlation with catchment characteristics. In choosing a conceptual model the *principle of parsimony* should be followed and the number of parameters should only be increased when there is clear advantage in doing so. These conclusions, based on an analytical approach, are confirmed by numerical experiments on both synthetic and natural data, which are described below.

6.1 USE OF MOMENT MATCHING

Once a conceptual model has been chosen for testing, the parameters for the conceptual model must be optimised i.e. must be chosen so as to simulate as closely as possible the actual unit hydrograph in some defined sense. In the present chapter attention will be concentrated on the *optimisation of model parameters by moment matching* i.e. by setting the required number of moments of the conceptual model equal to the corresponding moments of the derived unit hydrograph and solving the resulting equations for the unknown parameter values.

Optimisation of model parameters

This approach has the advantage that the moments of the unit hydrograph can be derived from the moments of the input and of the output through the relationship between the cumulants for a linear time-invariant system as given by equation (3.75). It has the second advantage that the moment relationship can be used to simplify the derivation for the moments or cumulants of conceptual models built up from simple elements in the manner described in the last two sections of Chapter 5.

The use of moment matching may be illustrated for the case of a cascade of linear reservoirs, which is one of the most popular conceptual models used to simulate the direct storm response. Since this is a two-parameter model we use the equations for the first and second moments and set these equal to the derived moments. The first moment is given by

$$nK = U_1'(h) = U_1'(y) - U_1'(x) \tag{6.1}$$

and the second moment by

$$nK^2 = U_2(h) = U_2(y) - U_2(x) \tag{6.2}$$

Once the first moment about the origin and second moment about the centre for the unit hydrograph have been determined from the corresponding moments of the effective precipitation and direct storm runoff, it is a simple matter to solve equations (6.1) and (6.2) in order to determine the values of n and K which are optimum in the moment matching sense.

Time–area–concentration curve

Where a conceptual model is based on the routing of a particular shape of *time–area–concentration curve* through a linear reservoir, the cumulants of the resulting conceptual model can be obtained by adding the cumulants of the geometrical figure representing the time–area–concentration curve and the cumulants of the linear reservoir.

Routed isosceles triangle

Thus, for the case of a *routed isosceles triangle* where the base of the triangle is given by T and the storage delay time of the linear reservoir by K, the cumulants of the resulting conceptual model are as follows. The first cumulant, which is equal to the first moment about the origin or lag, is given by

$$k_1 = U_1' = \frac{T}{2} + K \tag{6.3}$$

and the second cumulant or second moment about the centre by

$$k_2 = U_2 = \frac{T^2}{24} + K^2 \tag{6.4}$$

and the third cumulant or third moment about the centre by

$$k_3 = U_3 = 2K^3 \tag{6.5}$$

If the respective moments of this conceptual model are equated to the derived moment of an empirical unit hydrograph, then the value of the parameters that are optimal in the sense of moment matching can be evaluated.

In optimising the parameters of conceptual models by moment matching, it is necessary to have as many moments for the unit hydrograph as there are parameters to be optimised. The usual practice is to use the lower order moments for this purpose. This can be justified both by the fact that the estimates of the lower order moments are more accurate than those of higher order moments and also by the consideration that the order of a moment is equal to the power of the corresponding term in a polynomial expansion of the Fourier or Laplace transform. Reference was made earlier to the *convective-diffusion analogy*, which corresponds to a simplification of the St. Venant equations for unsteady flow with a free surface. This is a distributed model based on the *convective-diffusion equation*

Convective-diffusion analogy

$$D\frac{\partial^2 y}{\partial x^2} = a\frac{\partial y}{\partial x} + \frac{\partial y}{\partial t} \tag{6.6}$$

where D is the hydraulic diffusivity for the reach and a is a convective velocity. For a delta-function inflow at the upstream end of the reach, the impulse response at the downstream end is given by

$$h(x, t) = \frac{x}{\sqrt{4\pi Dt^3}} \exp\left[-\frac{(x - at)^2}{4Dt}\right] \tag{6.7}$$

which is a distributed model since the response is a function of the distance x from the upstream end. For a given length of channel, however, it can be considered as a lumped conceptual model with the impulse response

$$h(t, A, B) = \frac{A}{\sqrt{\pi t^3}} \exp\left[-\frac{(A - Bt)^2}{t}\right] \tag{6.8}$$

where $A = x/\sqrt{(4D)}$ and $B = a/\sqrt{(4D)}$ are two parameters to be determined. If moment matching is used, it can be shown that the value of A will be given by

$$A = \left(\frac{(U_1')^3}{U_2}\right)^{1/2} \tag{6.9}$$

and the value of B by

$$B = \left(\frac{U_1'}{U_2}\right)^{1/2} \tag{6.10}$$

These values are used in equation (6.8) in order to generate the impulse response.

6.2 EFFECT OF DATA ERRORS ON CONCEPTUAL MODELS

In Chapter 4 we discussed the performance of various methods of black-box analysis in the presence of errors in the data. It is interesting, therefore, to examine the performance of typical conceptual models under the same conditions. It will be recalled from Chapter 4 that the best method of direct matrix inversion was the Collins method, the most suitable method based

Table 6.1. Effect on unit hydrograph of 10% error in the data.

Method of identification	Mean absolute error as % of peak			
	Error-free	Systematic error	Random error	Mean for 10% error
Collins method	0.09×10^{-3}	5.8	27.7	16.8
Least squares	0.29×10^{-3}	6.6	21.5	14.1
Harmonic analysis ($N = 9$)	3.4	5.3	7.8	6.6
Meixner analysis ($N = 5$)	1.2	**4.8**	6.3	**5.6**
Nash cascade	2.8	6.0	**5.2**	**5.6**
Routed triangle	6.8	8.1	7.7	7.9
Diffusion analogy	7.0	8.0	7.4	7.7

on optimisation was the unconstrained least squares method, and that the best transformation methods were harmonic analysis and Meixner analysis. The results for these methods taken from Chapter 4 are reproduced in Table 6.1 together with the corresponding results for the three examples of two-parameter conceptual models discussed above. The parameter of the latter models were estimated by moment matching.

It is clear from Table 6.1 that all three conceptual models are more effective in filtering out random error than any of the algebraic methods of black-box analysis except those based on orthogonal functions. The success of the conceptual models in filtering out error in the derived unit hydrograph due to errors in the data may be explained by the fact that conceptual models automatically introduce constraints into the solution. Thus, all of the conceptual models automatically normalise the area of the unit hydrograph to unity, all of them produce only non-negative ordinates, and all of them produce unimodal shapes which are appropriate in the particular case under experimentation. It is important to remark in connection with the latter point that, if the actual unit hydrograph had

Bi-modal shape   a bi-modal shape, these particular conceptual models would not be able to compete with harmonic analysis or Meixner analysis. Accordingly the simple two-parameter conceptual models are able to compete successfully with complicated methods of black-box analysis in finding the true unit hydrograph in the presence of error at a level of 10%.

The conceptual models maintain their robust performance in the presence of higher levels of error, as indicated in Table 6.2, which shows the effect of the level of random error on the error in the unit hydrograph for various methods of identification. At a level of 15% the conceptual models continue to perform well and indeed perform better than harmonic analysis. The slow increase of the error in the case of the convective diffusion model might suggest that at higher levels of error it might prove more robust than the Nash cascade model and even than Meixner analysis.

Table 6.2. Effect of level of random error on unit hydrograph.

Method of identification	Mean absolute error as % of peak			
	Error-free data	5% error in the data	10% error in the data	15% error in the data
Collins method	0.09×10^{-3}	10.6	27.7	38.6
Least squares	0.29×10^{-3}	7.8	21.5	34.2
Harmonic analysis ($N = 9$)	3.4	5.1	7.8	14.2
Meixner analysis ($N = 5$)	1.1	**3.1**	**4.8**	**6.0**
Nash cascade	2.8	4.1	5.2	7.8
Routed triangle	6.8	7.0	7.7	9.1
Diffusion analogy	7.0	7.0	7.4	7.9

6.3 FITTING ONE-PARAMETER MODELS

Though unit hydrographs cannot in practice be satisfactorily represented by one-parameter conceptual models, it is remarkable the degree to which runoff can be reproduced by a one-parameter model. Conceptual models of the relationship between effective rainfall and direct storm runoff involving two or three parameters are of necessity more flexible in their ability to match measured data. However, in many cases the improvement obtained by using an additional parameter is much less than might be expected. This will be illustrated below, for the case of the data used by Sherman in his original paper on the unit hydrograph (Sherman, 1932a), and for the data used by Nash (1958) in the paper in which he first proposed the use of the cascade of equal linear reservoirs.

Even in the case of one-parameter conceptual models there is a wide choice available. We discuss below a number of conceptual models based on pure translation (i.e. on linear channels), on pure storage action (i.e. on linear reservoirs), and on the diffusion analogy.

The simplest one-parameter model based on pure translation is that of a *linear channel*, which displaces the inflow of its upstream end by a constant amount thus, shifting the inflow in time without a change of shape. The impulse response is a delta function centered at a time corresponding to the travel time of linear channel. Such a delta function has a first moment equal to the travel time but all its higher moments are zero. Thus the model based on a linear channel with upstream inflow will have a value of $s_2 = 0$ and a value of $s_3 = 0$. This model is shown as model 1 in Table 6.3, which lists the ten one-parameter conceptual models discussed in this section.

Linear channel with lateral inflow

It would, however, seem more appropriate in the case of catchment runoff (as opposed to a flood routing problem) to consider a *linear channel with lateral inflow*. If the inflow is taken as uniform along the length of the channel, then the instantaneous unit hydrograph would have the *shape*

Table 6.3. One-parameter conceptual models.

Model	Elements	Type of inflow	Shape factors	
			s_2	s_3
1	Linear channel	Upstream	0	0
2	Linear channel	Lateral, uniform	1/3	0
3	Linear channel	Lateral triangular (1 : 2)	1/6	0
4	Linear channel	Lateral triangular (1 : 3)	7/32	1/32
5	Linear reservoir	Upstream/lateral	1	2
6	2 reservoirs	Upstream	1/2	1/2
7	2 reservoirs	Lateral, uniform	7/9	10/9
8	3 reservoirs	Upstream	1/3	2/9
9	Diffusion reach	Upstream	∞	∞
10	Diffusion reach	Lateral, uniform	7/5	124/35

of a rectangle. In this case (model 2 in Table 6.3), the first moment would be given by $T/2$ and the second moment by $T^2/12$ thus giving a shape factor s_2 of $1/3$. Since the instantaneous unit hydrograph is symmetrical, the third moment and third shape factor are zero.

Recognising that most catchments are ovoid rather than rectangular in shape, we might replace this rectangular inflow by an *inflow in the shape of an isosceles triangle.* In this case the first moment is again given by $T/2$ and the second moment is $T^2/24$, thus giving a value of s_2 of $1/6$. The third moment and third shape factor would again be zero.

None of the three models mentioned above would be capable of reproducing the *skewness* which appears in most derived unit hydrographs. This of course could be overcome by using a *scalene triangle* rather than an isosceles in which the shape is kept fixed so that only one parameter is involved. In fact a triangle in which the *base length is three times the length of the rise* (model 4 in Table 6.3) was used by Sherman in his basic paper (Sherman, 1932a) and is illustrated in Figure 2.5.

If the one-parameter model is to be based on *storage*, the simplest model is that of a *single linear reservoir.* For this case (model 5) the value of s_2 as given by equation (5.25) is 1 and the value of s_3 as given by equation (5.26) above is 2. In the early studies of conceptual models carried out in Japan (Sato and Mikawa, 1956), the single linear reservoir was replaced by *two equal reservoirs in series* with the inflow into the upstream reservoir. If the number of reservoirs is kept constant in this fashion, it can be considered as a one-parameter model and for the case of two reservoirs both of the shape factors s_2 and s_3 will have the value of $1/2$ (model 6 in Table 6.3).

If one the other hand, we take *two equal linear reservoirs with lateral inflow* divided equally between them (model 7), then the shape factors are markedly different having the values of 7/9 and 10/9. If a cascade of *three*

Scalene triangle

Two equal linear reservoirs with lateral inflow

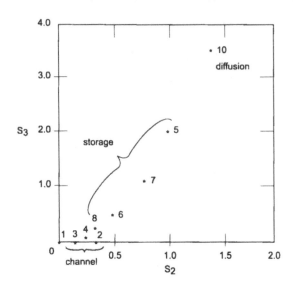

Figure 6.1. One-parameter conceptual models.

equal reservoirs is taken (model 8), then the values for the shape factor are 1/3 and 2/9. It must again be emphasised that unless the number of reservoirs is predetermined, these models cannot be considered as one-parameter models.

The diffusion analogy has been used as a conceptual model for surface flow, for flow in the unsaturated zone and for groundwater flow. If the model is one of pure diffusion without any convective term, then it can be classed as a one-parameter model. Where the inflow is taken at the upstream end of a diffusion element the first moment is infinite and all the higher moments are infinite. It can be shown that the shape factors s_2 and s_3 are also infinite. This means that the model cannot be fitted by equating the first moment of the model to the first moment of the data. However, the model corresponds to that represented by equation (6.8) above for the particular case where B is equal to zero. Accordingly the single parameter A can be determined from equation (6.9).

Diffusion reach with uniform lateral flow

Another one-parameter model (model 10 in Table 6.3) can be postulated on the basis of a *diffusion reach with uniform lateral flow*. In this case, which has been used in groundwater analysis and will be discussed in Chapter 7 (Kraijenhoff van de Leur et al., 1966), the moments are finite and the shape factor is given by 7/5 and 124/35.

A clear pattern is present in the values of the shape factors described above and listed in Table 6.3. The models based on translation give low values of the shape factors; those based on storage give intermediate values, and those based on diffusion give high values of the shape factor.

The models 1–10 listed in Table 6.3 are plotted on a shape factor diagram in Figure 6.1. Since they are all one-parameter models they plot as single points. All the above models have been included (along with a number of two-parameter and three-parameter models) in a *computer program* PICOMO, which is a special program for the identification of conceptual models (Dooge and O'Kane, 1977). Appendix A contains a detailed description of this program.

This program

(1) accepts sets of rainfall-runoff data;
(2) normalises the data;
(3) determines the moments of the normalised effective rainfall;
(4) determines the moments of the normalised direct runoff;
(5) computes the moments of the unit hydrograph by subtraction, and finally;
(6) computes the shape factors of this empirical unit hydrograph.

PICOMO contains Sheppard-type corrections in Activity 1 of the program, which apply when the system receives a truely pulsed input and a sampled output.

For each of the models included in the program, the parameter values are found by moment matching and the higher moments not used in the matching process are predicted. When the parameters have been determined the unit hydrograph is reconstituted and convoluted with the effective rainfall in order to generate the predicted runoff. The RMS error between the predicted and measured runoff is then determined.

For the *data of the Big Muddy river* (data set A) used by Sherman in his original paper (Sherman, 1932a) the peak for the unit hydrograph was 0.1337 and the time to peak was 16 hours. The *shape factors* of the derived unit hydrograph were $s_2 = 0.3776$ and $s_3 = 0.0335$. The Sheppard corrections have been used in generating these results. When they are not used s_2 is reduced by 0.5% and s_3 is increased by 1%, approximately. If we assumed that the inflow passed through the system unmodified (which could be considered as the *case of no model*) then the RMS error between this predicted outflow (equal to the inflow) and the measured outflow for this case would be 0.0659.

Table 6.4 shows the results of attempting to simulate Sherman's data by six of the one-parameter conceptual models described above. In each case the single parameter of the conceptual model would be found by equating its first moment to the first moment of the derived unit hydrograph. Table 6.4 shows the value for s_2 and s_3 of each of the models, which may be compared with the actual values of 0.3776 and 0.0335 given above.

Also shown in the table is the *RMS error* for each of the models and the predicted value of the *peak outflow* and the *time to peak*. It will be noted that the RMS error is least for the case of model number 2 where the model shape factor of $s_2 = 0.3333$ is closest to the empirical shape factor 0.3776. For this particular model the RMS difference between input and output has been reduced to 5% of its original value. In contrast for model

Table 6.4. One-parameter fitting of Sherman's Big Muddy data.

Model number	Shape factors		Predicted output		
	s_2	s_3	RMS error	q_p	t_p
1	0.0000	0.0000	0.0279	0.1897	17
2	**0.3333**	0.0000	**0.0037**	0.1301	16
4	0.2188	**0.0312**	0.0086	0.1469	**16**
5	1.0000	2.0000	0.0135	0.1245	15
6	0.5000	0.5000	0.0062	**0.1341**	**16**
9	∞	∞	0.0447	0.0418	18
Prototype	0.3776	0.0335		0.1337	16

Table 6.5. One-parameter fitting of Nash's Ashbrook data.

Model number	Shape factors		Predicted output		
	s_2	s_3	RMS error	q_p	t_p
1	0.0000	0.0000	0.0904	0.5384	11
2	0.3333	0.0000	0.0192	0.0608	14
4	0.2188	0.0312	0.0183	**0.1006**	9
5	1.0000	2.0000	0.0164	0.1015	4
6	**0.5000**	**0.5000**	**0.0069**	0.0876	7
9	∞	∞	0.0285	0.0304	8
Prototype	0.5511	0.6178		0.0994	5

Ashbrook catchment

RMS error

Time to peak

number 9, where the values of s_2 and s_3 are infinite, the RMS value is only reduced to 70% of its original value.

Similar results are obtained when an attempt is made to fit the *data of the Ashbrook catchment* (data set B) used by Nash in his first paper proposing the use of a cascade of equal linear reservoirs (Nash, 1958). In this case the *shape factors* derived for the unit hydrograph from the moments of the effective precipitation and the direct storm runoff were $s_2 = 0.5511$ and $s_3 = 0.6178$.

Table 6.5 shows the ability of the same six models used for Sherman's data to predict the derived unit hydrograph for Nash's Ashbrook data. As before this is measured by means of the *RMS error* between the predicted and observed output and the predicted peak and predicted time to peak. For no model (i.e. output equal to input) the RMS error between input and output was 0.1165, the *peak* of the derived unit hydrograph was 0.0994 and the *time to peak* of the derived unit hydrograph was 5 hours.

It will be seen from the table that for model 6 (two reservoirs in series with inflow into the upstream reservoir) the RMS error has been reduced from 0.1165 to 0.0069 i.e. to 6% of its original value. In contrast, for the case of model 1 (linear channel with upstream inflow) the fit is far from satisfactory and the RMS error is 0.0904 which is 80% of the original value.

The two examples given above illustrate the power of a one-parameter model to represent data, provided we can select an appropriate one-parameter model. It will be noted that in each of the above examples the one-parameter model, which gave the best performance in terms of RMS error between predicted and observed output, was the model whose value of s_2 was closest to the estimated value of s_2 for the derived unit hydrograph. It is important to note that in this case the criterion for judging the accuracy of the model (the RMS error) was different from that on which the optimisation of a single parameter and the selection of the appropriate model was based (i.e. moment matching).

6.4 FITTING TWO- AND THREE-PARAMETER MODELS

We now examine what improvement can be gained by the use of two-parameter models. There is naturally a wide choice available. The two-parameter models included in the computer program PICOMO are listed in Table 6.6.

Any shape of lateral inflow to a linear channel that involves two parameters will provide a two-parameter conceptual model of direct storm runoff. Model 11 in Table 6.6 involves a triangular inflow of length T with the peak at the point aT. Models 3 and 4 in Table 6.3 are obviously special cases of model 11. As remarked previously the unit hydrograph described by Sherman in his original paper (Sherman, 1932a) was a triangular unit hydrograph with the base three times the time of rise i.e. with the value of $a = 1/3$. Similarly the shape of the unit hydrograph used in the Flood Studies Report published in the United Kingdom (NERC, 1975) uses a triangular unit hydrograph with a value of a approximately equal to 0.4.

Translation   A two-parameter model can always be obtained by combining any one-parameter model based on *translation* (i.e. models 1 to 4 in Table 6.3) with a single linear reservoir. The two-parameter models corresponding

Table 6.6. Two-parameter conceptual models.

Model number	Elements	Type of inflow
11	Channel	Lateral, triangular (a : 1)
12	Channel plus reservoir	Upstream
13	Channel plus reservoir	Lateral, uniform
14	Channel plus reservoir	Lateral, triangular (1 : 2)
15	Channel plus reservoir	Lateral, triangular (1 : 3)
16	*n* equal reservoirs	Upstream
17	2 unequal reservoirs	Upstream
18	*n* equal reservoirs	Lateral, uniform
19	2 unequal reservoirs	Lateral, non-uniform
20	Convective diffusive	Upstream

to models 1 to 4 in Table 6.3 are listed as models 12 to 15 in Table 6.6. The moments (or cumulants) of the resulting models are obtained by adding the moments (or cumulants) of model 5 in Table 6.3 to the moments (or cumulants) of the appropriate translation model.

Storage

It is also easy to construct two-parameter models based solely on *storage*. Models 5, 6 and 8 in Table 6.3 represent the cases of an upstream inflow into a cascade of *one, two and three equal reservoirs* respectively. These are all special cases of the Nash cascade which consist of a series of *n* equal linear reservoirs (model 16 in Table 6.6). Alternatively model 6 in Table 6.3 which is a one-parameter model based on two-equal reservoirs each with a delay time K can be modified to give a two-parameter model based on two *reservoirs with unequal delay times* (K_1 and K_2) placed in series thus giving model 17 in Table 6.6. Model 7 in Table 6.3 i.e. two equal reservoirs with uniform later inflow can be modified in a number of ways.

Lateral inflow

The uniformity of *lateral inflow* can be retained and the length of the cascade used as a second parameter thus giving model 18 in Table 6.6. Alternatively the length of the cascade could be retained at two and the lateral inflow into each reservoir varied, thus giving model 19 in Table 6.6.

Finally the models based on *diffusion* can be modified by the introduction of a *convective term* thus giving model 20 in Table 6.6. This model has already been referred to and its lumped form is given by equation (6.8) above. Model 14 (*routed isosceles triangle*), model 16 (*cascade with upstream inflow*) and model 20 (*convective-diffusion analogy*) have already been compared on a shape factor diagram in Figure 5.2, and again in Figure 6.2. They plot relatively close to one another, in spite of the fact that the conceptual models are based on differing concepts of translation, storage and diffusion.

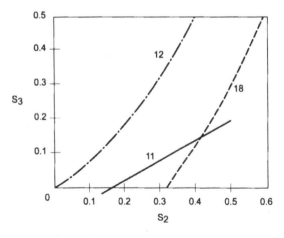

11 Channel with triangular inflow
12 Channel and reservoir
18 Cascade with lateral inflow

Figure 6.2. Shape factor comparison of conceptual models.

Table 6.7. Two-parameter fitting of Sherman's Big Muddy data.

Model number	Shape factors		Predicted output		
	s_2	s_3	RMS error	q_p	t_p
11	0.3776	**0.1260**	**0.0036**	0.1381	14
12	0.3776	0.4623	0.0085	0.1412	16
13	0.3776	0.3478	0.0070	0.1435	16
14	0.3776	0.4118	0.0083	0.1464	16
16	0.3776	0.2837	0.0061	0.1405	16
17	0.3776	0.9543	0.0086	**0.1305**	15
19	0.3776	0.3479	0.0074	0.1407	16
20	0.3776	0.4256	0.0083	0.1461	16
Prototype	0.3776	0.0335	–	0.1337	16

Further comparison of two-parameter conceptual models is shown in Figure 6.2. The conceptual models shown are model 11 (a linear channel with lateral inflow in the shape of a scalene triangle), model 12 (upstream inflow into a linear channel followed by a linear reservoir) and model 18 (a cascade of equal linear reservoirs with equal lateral inflow). It can be seen in this case that the curves plot well apart on a shape factor diagram. Accordingly the models afford a degree of flexibility in matching the plotting of derived unit hydrographs.

Two-parameter models

The *fitting of certain two-parameter models* to the data of Sherman is shown in Table 6.7. Since we have two parameters at our disposal both the scale factor and the s_2 shape factor can be fixed in this case. Accordingly the value of s_2 of the derived unit hydrograph of 0.3776 will be matched exactly by each of the two-parameter models. It will be noted from Table 6.7 that the RMS error is least (and the peak is most closely approximated) by model 11 for which the value of s_3 is closest to the derived value of 0.0335. Model 11 is the conceptual model based on taking the shape of the unit hydrograph as a scalene triangle.

It is also worthy of note that the RMS error does not vary widely for the two-parameter models studied. The RMS error between the predicted and observed output ranges from 5% to 13% of the initial RMS error. It is also noteworthy that the best two-parameter models when compared with the best one-parameter model only shows a reduction of the RMS error from 0.0037 to 0.0036.

Similar results are obtained when the two-parameter models are applied to the data for the *Ashbrook catchment* (Nash, 1958) and are shown in Table 6.8. All of the two-parameter models give fairly similar levels of performance, the RMS error varying from 6% to 10% of the original RMS error for no model (i.e. output equal to input). Again the value of s_2 is the same in all models and the best performance is given by model 16 whose value of s_3 is closest to the derived value of s_3 of 0.1678. This model is the Nash cascade of n equal linear reservoirs with upstream inflow.

Table 6.8. Two-parameter fitting of Nash's Ashbrook data.

Model number	Shape factors		Predicted output		
	s_2	s_3	RMS error	q_p	t_p
11	0.5511	0.1998	–		–
12	0.5511	0.8182	0.0089	0.1303	6
13	0.5511	0.7628	0.0071	0.1052	7
14	0.5511	0.7922	0.0072	0.1113	7
16	0.5511	**0.6074**	**0.0068**	0.0864	6
17	0.5511	1.6082	0.0113	**0.0985**	**5**
19	0.5511	0.5826	0.0078	0.0835	6
20	0.5511	0.9111	0.0082	0.1117	7
Prototype	0.5511	0.6178		0.0994	5

For this data also, the RMS error is only reduced slightly when we move from the best one-parameter model to the best two-parameter model, being 0.0069 for the model 6 (two reservoirs in series) and 0.0068 for model 16 (a Nash cascade). The small improvement is explicable in this case. The optimum value of n for the two-parameter cascade model is 1.8, which is close to the fixed value of 2 in the one-parameter model 6.

The results discussed above would suggest that there would be very little advantage in extending the number of parameters to three in the fitting of the two sets of data. However, a discussion of this step is included here for the sake of completeness.

A very large number of three-parameter models can be synthesized in an attempt to simulate the operation of the direct storm runoff or any other component of catchment response. A two-parameter model of a channel with lateral inflow in the shape of a scalene triangle (model 11 in Table 6.7) can be combined with a single linear reservoir to give a conceptual model based on a *routed scalene triangle* (model 21). Similarly two-parameter model 12 (linear channel plus linear reservoir) can be combined with two-parameter model 16 (n equal reservoirs) to give a three-parameter conceptual model based on upstream inflow to a channel and a cascade of equal linear reservoirs in series i.e. a *lagged Nash cascade* (model 22). Similarly model 17 in Table 6.6 (two unequal reservoirs with upstream inflow) can be given an additional parameter either by adding a third unequal reservoir (model 23) or by changing from upstream inflow to non-uniform lateral inflow (model 24).

Routed scalene triangle

Lagged Nash cascade

When moment matching is used to apply conceptual models to field data, it frequently gives rise to a negative or complex value for a physic-ally based parameter. If these unrealistic values are not accepted and the particular parameter set equal to zero, the three-parameter model is in fact reduced to a two-parameter model.

For the case of the Big Muddy River data, the PICOMO program, when tested for models 22, 23 and 24 found no realistic parameter values.

For the Ashbrook data (Nash, 1958), no realistic values were obtained for model 23, but acceptable values for all three parameters were obtained in the case of models 22 and model 24. For the case of model 22 (a linear channel followed by n equal linear reservoirs) the RMS error was 0.0064 compared with 0.0068 for the best two-parameter model (n equal linear reservoirs without the channel). For the case of model 24 (two unequal reservoirs with non-uniform lateral inflow) the RMS error between the predicted and observed outputs was 0.0071. This is not as good as the best two-parameter model, but better than either of the 2 two-parameter models tested, which are special cases of model 24. These are models 17 (two-unequal reservoirs with upstream inflow which had a RMS error of 0.0113) and model 19 (two equal reservoirs with non-uniform lateral inflow which had a RMS error of 0.0078).

The above *results may be summarised* by saying that for the data examined

(a) the original RMS error between input and output can be reduced to less than 10% of its original value by means of a one-parameter model, if one can be found with a value of s_2 close to that of the derived unit hydrograph;

(b) the use of a two-parameter model guarantees that the value s_2 will be matched and that the RMS error will be an order of magnitude less than the original value; and finally

(c) the addition of a third parameter brings little improvement and may lead to unrealistic parameter values.

Relative efficiency

The *relative efficiency* of a suitable one-parameter model, if found, and the relative inefficiency of additional parameters are illustrated for the case of *Sherman's Big Muddy data* in Table 6.9. In this case, none of the three-parameter models tried, gave realistic parameter values, and consequently, there is no improvement over the two-parameters results. The results for the best models with one, two and three parameters for *Nash's Ashbrook data* are shown in Table 6.10. In this case realistic parameters were obtained for two of the three-parameter models. It can be seen from the table, however, that the improvement by the addition of the second and the third parameter are not substantial.

Table 6.9. Best models for Sherman's Big Muddy River data.

Number of parameters	Best model		RMS error	
	Model number	Description	Value	%
0	–	Outflow = inflow	0.0659	100.0
1	2	Rectangle	0.0037	5.6
2	11	Triangle (a : 1)	0.0036	5.5
3	–	–		

Replacing a two-parameter model by a three-parameter model may give rise to unrealistic parameter values. This is analogous to the case in black-box analysis where lengthening the series gives worse results, because the added flexibility results in the model fitting itself to the noise rather than to the underlying signal.

PICOMO program

The PICOMO program (on which the above comparisons are based) attempts to fit 17 of the 24 one-, two- and three-parameter. The models form a family, the structure of which can be presented by the directed graph shown in Figure 6.3. See any text on the theory of graphs, such as Kaufmann (1968). Each vertex X_i corresponds to a model. Pairs of vertices X_i and X_j are connected by arcs (X_i, X_j) in order to indicate the relation that the model at the terminal extremity of an arc X_j contains as a special case the model at the initial extremity X_i. For example, model 22

Table 6.10. Best models for Nash's Ashbrook Catchment data.

Number of parameters	Best model		RMS error	
	Model number	Description	Value	%
0		Outflow = inflow	0.1165	100.0
1	6	2 equal reservoirs	0.0069	5.9
2	16	n equal reservoirs	0.0067	5.8
3	22	Channel plus n reservoirs	0.0064	5.5

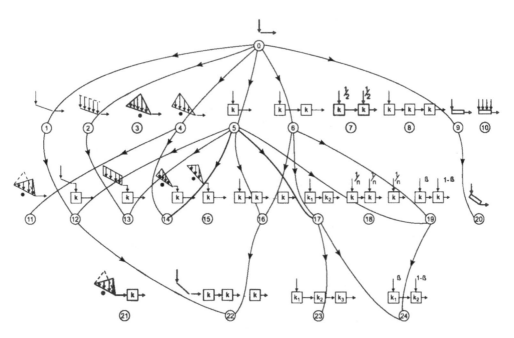

Figure 6.3. The model inclusion graph (Dooge and O'Kane, 1977).

the "lagged cascade of n equal reservoirs", contains model 12: "lag and route", and model 16: "Nash cascade" as special cases. Hence we define the arcs (X_{12}, X_{22}) and (X_{16}, X_{22}) in order to represent this inclusion. Models 12 and 16, in their turn, contain models 1, 5 and 6 as special cases. Hence, we define the arcs (X_1, X_{12}), (X_5, X_{12}), (X_5, X_{16}), (X_6, X_{16}), and so on.

Model inclusion

The binary relation of model inclusion is

(a) strictly anti-symmetric, i.e. if X_j includes X_i as a special case then X_i cannot include X_j as a special case; and

(b) transitive, i.e. if model X_j includes X_i, and X_k includes X_j, then X_k also includes X_i as a special case.

Hence the relation of model inclusion always defines a *strict* ordering of the models and the graph showing this will have no circuits. The ordering is *partial*, not total, since models with the same number of parameters cannot be related by inclusion.

The strict ordering of the models by inclusion is not shown in its entirety in Figure 6.3, e.g. (X_{22}, X_5) is not shown since it is implied by (X_{22}, X_{16}) and (X_{16}, X_5). This is done for clarity. In addition, only those arcs which relate the 17 models in the program are shown. Hence Figure 6.3 is a partial graph obtained by deleting arcs from the full graph, which represents the strict-order relation defined by model inclusion on the 24 models considered above. Model 0 is the model whose outflow is equal to its inflow and has no non-zero parameters. Since it is included as a special case in every other model and has no special case itself, its vertex X_0 is the *minorant* of the graph in Figure 6.3.

This strict ordering will be used subsequently to display

(a) the consistency of measures of goodness of fit other than moments; and

(b) the improvement in measures of fit with increasing numbers of parameters.

One can then attempt to trade extra parameters against greater model accuracy.

If the ability to match moments were a perfect predictor of RMS error, which it is not, then a plot of RMS error on the model inclusion graph would show a strict ordering of the models with respect to RMS error. Figure 6.4 shows the case for the Big Muddy River data (data set A). The filled circular symbols on three of the arcs show where violations of this order occur. Figure 6.5 shows the case for the Ashbrook Catchment data (data set B). The filled circular symbols show that there are two direct violations, and one indirect violation in this case; the offending arc (X_6, X_{24}) is not shown. The strict ordering of the models is reproduced in all other cases. Clearly, this type of systematic analysis can be repeated for any other measure of fit, e.g. time to peak.

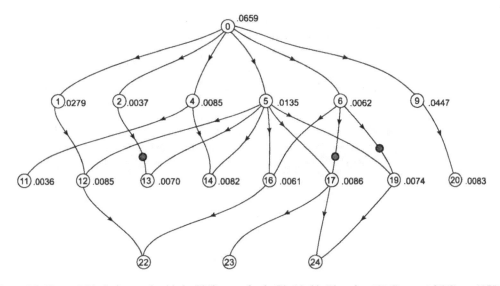

Figure 6.4. The model inclusion graph with the RMS errors for the Big Muddy River data (A) (Dooge and O'Kane, 1977).

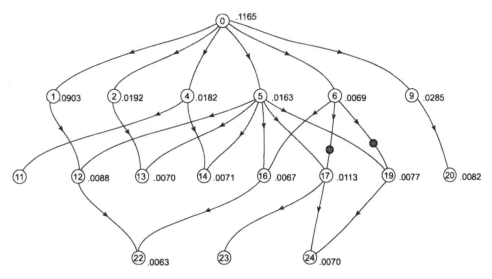

Figure 6.5. The model inclusion graph with the RMS errors for the Ashbrook Catchment data (B) (Dooge and O'Kane, 1977).

The model inclusion graph only displays a partial though strict ordering of the models. Hence a further comparison of RMS error between models with the same number of parameters is necessary in order to attempt a total ordering of the models. This in turn can be represented by another graph. In data set A the best one-parameter model: 2. Rectangle, is better than all two-parameter models with the exception of model: 11. Scalene triangle.

None of the three-parameter models produced realistic parameters. The sensitivity of these results to changes in the number of active rainfall ordinates has not been investigated. In data set B, model: 6, two equal reservoirs with upstream inflow, is the best one-parameter model and is surpassed only by the two-parameter model: 16, Nash cascade, and by the three-parameter model: 22, the lagged Nash cascade.

Law of diminishing returns

In both cases a law of diminishing returns appears to hold for the models considered. The RMS of the zero model is merely the RMS difference between inflow and outflow. The inclusion of an appropriate one-parameter model reduces this by at least an order of magnitude. However, the addition of further parameters produces a marginal decrease in RMS error. In addition physically unrealistic values of the parameters occur more frequently.

6.5 REGIONAL ANALYSIS OF DATA

Synthetic unit hydrographs

General synthetic scheme

It will be recalled from Section 5.1 that conceptual models first arose in the context of *synthetic unit hydrographs*. Unit hydrographs can be derived for the gauged catchments in a region and made the basis of a synthetic unit hydrograph for the ungauged catchments in the same region. For a *general synthetic scheme*, it is necessary to determine

(a) the degree of complexity (i.e. the number of parameters) required in the conceptual model;
(b) the particular model of this degree of complexity which best represents the gauged catchments; and
(c) the correlation between some parameters of the chosen model and suitable catchment parameters.

The moments of the individual unit hydrographs, which can be determined from the moments of effective rainfall and of storm runoff, can be used systematically as the basis for a general synthetic scheme incorporating all three phases listed above. Such a general synthetic scheme was first suggested by Nash (1959, 1960). He proposed that *the derived moments of the unit hydrograph for the gauged catchments should be correlated with one another and with the catchment characteristics*, to determine the number of degrees of freedom inherent in the response of the catchments when operating on precipitation excess to produce direct flood runoff. The number of degrees of freedom determines the number of parameters needed in the synthetic unit hydrograph. He suggested that the dimensionless moments of the actual unit hydrograph should be plotted against one another thus producing what has been called a *shape-factor diagram* in Section 6.3 and 6.4 above. If the plotted points clustered around a single point then a one parameter model would be indicated. If the points fell close to a line and this line could be identified with a particular conceptual model then his two-parameter conceptual model could be used. If the plotted points filled a region, an attempt could be made to find a three-parameter model, which would cover the same region.

It was suggested by Dooge (1961) in the discussion of Nash's paper that this scheme could, with advantage, be modified. *The moments should be correlated among themselves, rather than with the catchment characteristics*, in order to determine the number of degrees of freedom. Thus, in a two-parameter system, the third moment would be completely determined, once the first and second moments were known. Similarly, in a three-parameter system, the fourth moment would be known, once the first, second and third moments were known.

If the moments are made dimensionless by using the first moment as a scaling factor, then the criterion for a two-parameter model would be that the third dimensionless moment (i.e. the third dimensionless cumulant) would be completely determined by the second dimensionless moment (or cumulant). Similarly the criterion for a three-parameter system would be that the dimensionless fourth moment (or cumulant) would be completely determined by the second and the third dimensionless moments.

The remainder of the modified general synthetic scheme, which is shown in Figure 6.6, is essentially the same as that for Nash's original proposal. The shape factors of the derived unit hydrographs can be used to choose the most appropriate conceptual model with the appropriate number of parameters. The unit hydrograph parameters for the chosen model can be correlated with catchment characteristics on the basis of the moment of the derived unit hydrographs for the gauged catchments. *Ungauged catchment*   *To obtain a synthetic unit hydrograph for an ungauged catchment*, the unit hydrograph parameters are obtained from the correlation with catchment characteristics and then used with the selected model to generate the required unit hydrograph.

In his paper, Nash (1959) analysed the *data for 90 storms on 30 catchments* in Great Britain. Dooge (1961) calculated the coefficient of

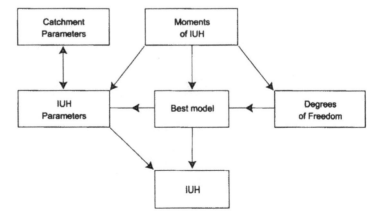

Figure 6.6. The general synthetic scheme (Dooge, 1961).

multiple correlation of s_3 with s_2 for Nash's data as 0.717. This indicated that only 50% of the variation in the third dimensionless moment was accounted for by variations in the dimensionless second moment (s_2). Hence, a two-parameter model would not be highly efficient as a basis for simulation.

However, the coefficient of multiple correlation between the dimensionless fourth moment and the two lower dimensionless moments was found to be 0.93, thus indicating that the variance in the fourth dimensionless moment was accounted for by the variance in the lower dimensionless moments of the extent of almost 90%. Considering the basic nature of Nash's data (which were normal river observations rather than research readings) it was a very high correlation and indicated that the three-parameter model would probably give a satisfactory simulation of the data obtained for all unit hydrographs in the region of Great Britain.

Nash's data are plotted on a (s_3, s_2) shape factor diagram in Figure 6.7. It can be seen from this plotting that the points define a region rather than a line in the shape factor diagram thus confirming the result of the correlation analysis. A two-parameter model would hardly have been adequate to represent all the unit hydrographs. At least three parameters are necessary to achieve this end.

The limiting forms of two-parameter models discussed in Section 5.4 are also drawn in Figure 6.7. These derived unit hydrographs fall within the limits, which apply to the general model of a cascade of linear reservoirs (not necessarily equal) with any distribution of positive lateral inflow. It is also noteworthy that the line for the Nash cascade plots in a central position.

There remains the problem of correlating the required number of unit hydrograph parameters. The number of relationships necessary,

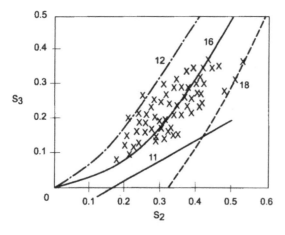

11 Channel with triangular inflow
12 Channel and reservoir
16 Cascade with upstream inflow
18 Cascade with lateral inflow

Figure 6.7. Shape factor plotting of regional data.

*Unit hydrograph
parameters*

corresponds to the number of parameters required for the conceptual model. Difficulties arise both in regard to the *choice of unit hydrograph parameters* and to the *choice of the catchment parameters*. Parameters relating directly to the shape of the derived unit hydrograph, usually belong to one of the three types

(1) time parameters;
(2) peak discharge parameters; and
(3) recession parameters.

Most of the early work on synthetic unit hydrographs used parameters of the derived unit hydrographs such as the time to peak and the peak discharge.

The most important *time parameters* used in synthetic unit hydrographs are shown in Figure 6.8. In this figure, t_0 is used to denote the duration of precipitation excess, which is assumed to occur at a uniform intensity over this unit period.

Common time parameters based only on the *outflow hydrograph* that have been used in synthetic unit hydrograph studies to characterise the outflow hydrograph are

(1) the time of rise of the unit hydrograph (t_R), i.e. the time from the beginning of runoff to the time of peak discharge;
(2) the time of virtual inflow (T), i.e. the time from the beginning of runoff to the point of cessation of recharge to groundwater storage; or
(3) the base length of the unit hydrograph i.e. the total runoff time (B).

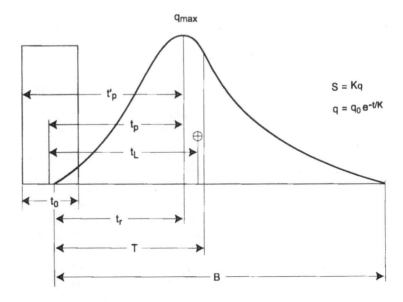

Figure 6.8. Typical unit
hydrograph parameters.

Time parameters

The common time parameters used to connect the *precipitation excess* and the hydrograph of *direct runoff* include

(1) the lag time or time from the centre of mass of precipitation excess to the centre of mass of direct storm runoff (t_L);
(2) the lag to peak time or the time from the centre of mass of effective precipitation to peak of the hydrograph (t_p); or
(3) the time to peak, i.e. the interval between the start of the rain, and the peak of the outflow hydrograph (t'_p).

One of the most important factors in surface water hydrology is the delay imposed on the precipitation excess by the action of the catchment. If the parameter representing this delay is to be useful for correlation studies, it should be independent of the intensity and duration of rainfall. In the case of a linear system — and the unit hydrograph method assumes the system under study to be linear — the time parameters listed above are all independent of the intensity of precipitation excess, but only the *lag time* (t_L) has the property of being *independent of both the intensity and the duration* of the precipitation excess. Accordingly, with the hindsight given by the systems approach we can say that only the lag time should be used as the duration parameter studied under synthetic unit hydrographs.

Discharge parameters

In regard to *discharge parameters*, the peak discharge q_{max} is almost invariably used when such a parameter is required. Another parameter, which can be estimated for a derived unit hydrograph, is the time parameter K that characterises the recession of the unit hydrograph when this recession is of declining exponential form. In such cases, the unit hydrograph may be considered as having being routed through a linear reservoir whose storage delay time is K. If the recession can be represented in this form, a plotting of the logarithm of the discharge against time will give a straight line and the value K can be estimated from the slope of this line. Alternatively, the value K may be determined at any point on the recession curve by dividing the remaining outflow after that point by the ordinate of outflow at the point. Other parameters used to characterise the unit hydrograph are the values of W-50 and W-75 which are defined as the width of the unit hydrograph for ordinates of 50% and 75% respectively of the peak value.

As indicated already, Nash (1958, 1959, 1960) suggests the use of the *statistical moments* of the instantaneous unit hydrograph as the determining parameters both for the identification of the unit hydrograph and for the correlation with catchment characteristics. The first moment about the origin of the instantaneous unit hydrograph is identical to the lag time defined above and recommended as an appropriate delay parameter. The second and third moments have the advantage over parameters, such as the peak discharge, that they are based on all of the ordinates of the unit hydrograph and not on single points. They are therefore more stable in the presence of errors of measurement or of derivation.

Instead of correlating characteristics of the unit hydrograph with catchment characteristics, this correlation could be based on the value of the parameters of the conceptual model, which are chosen for the fitting of the data. Since these are chosen by moment matching, or some other process, which takes the whole of the response curve into account, they have the stability characteristics spoken of above, in regard to the statistical moments.

Catchment characteristics

Scale factor

The *choice of catchment characteristics* for use in a correlation process also gives rise to difficulty. As might be expected, all synthetic unit hydrograph procedures involve a *scale factor* but a variety of scale factors are used in practice:

(1) the area of the catchment itself (A);
(2) the length of the main channel;
(3) the length of the highest order of stream (L);
(4) the length to the centre of area of the catchment (L_{ca}); or
(5) for a small catchment the length of overland flow (L_o).

A review of synthetic unit hydrograph procedures reveals a *slope* as the second most frequently used catchment characteristic. Since slope varies throughout a watershed, a standard definition of some representative slope is required. The slope parameters most often used are the average slope of the main channel or some average slope of the ground surface. The measurement of average slope parameters usually involves tedious computations (Strahler 1964; Clarke 1966). Although area (or stream length) and channel (or ground) slope have been used almost universally, there is no agreement about the remaining catchment characteristics which might be used.

The *shape of the catchment* must have some effect, but there is a wide variety of shape factors of choose from (form factors, circularity ratios, elongation ratios, leminiscate ratios, etc) and the lack of uniformity is not surprising. If there is considerable storage in the catchment, the effect of shape on the unit hydrograph pattern may not be very marked. Another factor, which must affect the hydrograph, is the *stream pattern*, which may be represented by *drainage density* or *stream frequency* or some such parameter. Although parameters representing the mean characteristics must have primary influence, the *variation in certain characteristics* from part to part of the catchment will give rise to secondary parameters whose effect may not be negligible. Thus, having taken area and slope into account, the third most important parameter may well be variation of length or of slope rather than a new parameter describing shape or drainage density.

It must be stressed that what is required in a correlation for a unit hydrograph synthesis is not necessarily a correlation with individual catchment characteristics, but rather with independent catchment parameters. These may be made up from a number of characteristics in the same way as the Froude number and the Reynolds number in hydraulic modelling are made up from a number of hydraulic characteristics. The choice of catchment characteristics for correlation with unit hydrograph parameters will remain a subjective matter, until we have a deeper knowledge of the *morphology of natural catchments*. The latter is a vital subject for the progress of hydrology. Despite the advance made by the introduction of the concept of the geomorphological unit hydrograph (GUH) by Rodriguez-Iturbe and Valdes (1979), the progress in relating this concept to hydrologic practices in the last two decades has been disappointing. The close approximation of the shape of the GUH to the IUH of the Nash Cascade of equal linear reservoirs has been noted (Chuta and Dooge, 1991) on the basis of 1100 Monte Carlo simulations of the GUH for a third-order catchment. This result was later generalised to second-order, fourth-order and fifth- and sixth-order catchments by Shamseldin and Nash (1998).

A review of the various studies of the correlation of catchment unit hydrograph parameters with catchment characteristics reveals that, in most cases, it is possible to get a reliable estimate of the lag or similar time parameter, but not to obtain a second or third relationship, which is necessary in the case of two or three parameter conceptual models. Probably *Best procedure* the *best procedure in any new study* is to attempt to correlate the lag of the derived unit hydrograph with the area (or some length characteristic) and the slope by means of a relationship such as

$$t_L = C\frac{(A)^d}{(S)^b} \tag{6.11}$$

where C is an empirical constant. As indicated below, a number of types of relationships might be sought in order to supplement this basic relationship, but previous workers in the field have not been successful in finding reliable correlations other than for the lag of the catchment.

One of the most commonly used early methods for synthetic unit hydrographs is that due to Snyder (1938) which is based on data from *Appalachians* *twenty catchments in the Appalachians*. He took as the basic unit hydrograph parameter the *lag time to peak* (t_p) defined as the time in hours between the centre of rainfall and the peak of the unit hydrograph and took as the basic catchment characteristic the product of the *length of the main channel* in miles (L) and the *length from the outlet to the centre of area of the catchment* in miles (L_{ca}).

He suggested that the unit hydrograph parameter and the catchment characteristic could be connected by the equation

$$t_p = c_t(LL_{ca})^{0.3} \tag{6.12}$$

where c_t is an empirical parameter. Having determined the time to peak of the unit hydrograph, Snyder assumed that the peak of the unit hydrograph could be determined from a second relationship

$$q_{max} = 640\frac{c_p}{t_p} \tag{6.13}$$

where q_{max} is the unit hydrograph peak in cubic feet per second per square mile, t_p is the time to peak in hours given in equation (6.12), and c_p is a coefficient that takes account of the flood wave storage in the catchment. For the catchments, which he studied in the Appalachians, Snyder found c_t to vary between 1.8 and 2.2 and c_p to vary between 0.56 and 0.69.

Snyder in his original paper (1938) published a diagram for deriving the ordinates of the twenty-four hour distribution graph; but this was not adopted by later workers who used his basic method. A number of subsequent workers used Snyder's form of relationship between *lag time to peak* and the *catchment length*. Linsley and others (1949) found the value of the coefficient c_t to vary from 0.7 and 1.0 for catchments in the Sierra Nevada. The Corps of Engineers (1963) found the same parameter to vary from 0.4 in Southern California to 0.8 for catchments bordering the Gulf of Mexico. The Corps of Engineers investigations also indicated that the value of c_p could vary from 0.31 in the Gulf of Mexico to 0.94 in Southern California.

British catchments

Nash (1960) sought direct correlation with catchment characteristics of the moments of the unit hydrographs, which he derived. On the basis of *ninety storms* on *thirty British catchments* (with area varying from 4.8 to 859 square miles), Nash (1960) derived the relationship

$$t_L = U_1' = 27.6 \left(\frac{A}{S}\right)^{0.3} \tag{6.14}$$

where U_1' is the first moment in hours, A is the area in square miles, and S is the overland slope in parts per thousand. Before adopting this relationship, Nash had tried the regression of the first moment on various combinations of nine catchment characteristics. The coefficient of multiple correlation (R) for the relationship given in equation (6.14) was 0.90, which indicates that it is a reliable correlation.

When the dimensionless second moment was correlated against catchment characteristics the best result obtained was

$$s_2 = \frac{U_2}{(U_1')^2} = \frac{0.41}{(L)^{0.1}} \tag{6.15}$$

where s_2 is the dimensionless second moment about the centre and L is the length of the longest stream from the outlet to the catchment boundary in miles. In this second regression relationship, the coefficient of multiple correlation (R) was only 0.5.

Irish catchments

As an example of the use of parameters with conceptual models for correlation, we can take the case of the routed isosceles triangle applied to drained Irish catchments by O'Kelly (1955). For this particular conceptual model the two parameters are the base of the isosceles triangle (T) and the storage delay time of the linear reservoir (K). Based on O'Kelly's data, Dooge (1955) derived the relationship

$$T = 2.58\frac{(A)^{0.41}}{(S)^{0.70}} \tag{6.16}$$

where T is the base length of the triangular inflow in hours, A is the catchment area in square miles and S is the slope in parts per ten thousand; and he also derived the second relationship

$$K = 100.5\frac{(A)^{0.23}}{(S)^{0.70}} \tag{6.17}$$

where K is the storage delay time in hours.

The National Authorities in many countries have established standard procedures for flood studies. The discussion of these studies is outside the scope of this book. In cases where catchment behaviour is affected by *human activity*, it is necessary to modify the unit hydrograph derivation and the regional analysis. For the case of a single catchment it is necessary to determine parameter values at different levels of a given factor of human influence and to attempt to find a correlation between the values of the parameters at each level of influence. In some cases, it may not be possible to obtain data on a single catchment at different stages of development and accordingly we must again resort to a regional analysis in which the catchments have different characteristics and also different levels of development. Much less work has been done in this area than in the general field of synthetic unit hydrographs, but it is becoming of interest because of the necessity of dealing with the problem of *urbanisation* and other changes in *land use*.

Urbanisation

Land use

CHAPTER 7

Simple Models of Subsurface Flow

7.1 FLOW THROUGH POROUS MEDIA

In Chapters 5 and 6 we have been concerned with the black box analysis and the simulation by conceptual models of the direct storm response, i.e. of the quick return portion of the catchment response to precipitation. The difficulties that arise in the unit hydrograph approach concerning the baseflow and the reduction of precipitation to effective precipitation, arise from the fact that these processes are usually carried out without even postulating a crude model of what is happening in relation to *soil moisture* and *groundwater*. Even the crudest model of subsurface flow would be an improvement on the classical arbitrary procedures for baseflow separation and computation of effective precipitation used in applied hydrology. It is desirable, therefore, for the study of floods as well as of low flow to consider the slower response, which can be loosely identified with the passage of precipitation through the *unsaturated zone* and through the *groundwater reservoir*. In other words, it is necessary to look at the remaining parts of the simplified catchment model given in Figure 2.3 (see page 19). We approached the question of prediction of the direct storm response through the black-box approach in Chapter 4 and then considered the use of conceptual models as a development of this particular approach in Chapter 5. In the case of *subsurface flow*, we will take the alternative approach of considering the equations of flow based on physical principles, simplifying the equations that govern the phenomena of *infiltration* and *groundwater flow* and finally developing lumped conceptual models based on these simplified equations.

Groundwater

The basic physical principles governing subsurface flow can be found in the appropriate chapters of such references as Muskat (1937), Polubarinova-Kochina (1952), Luthin (1957), Harr (1962), De Wiest (1966), Bear and others (1968), Childs (1969), Eagleson (1969), Bear (1972), and others. The movement of water in a saturated porous medium takes place under the action of a potential difference in accordance with the general form of *Darcy's Law*.

Darcy's Law

$$V = -K \operatorname{grad} (\phi) \qquad (7.1)$$

where V is the rate of flow per unit area, K is the *hydraulic conductivity* of the porous medium and ϕ is the *hydraulic head* or potential. If we neglect the effects of temperature and osmotic pressure, the potential will be equal to the *piezometric head* i.e. the sum of the pressure head and the elevation:

$$\phi = h = \frac{p}{\gamma} + z = -S + z \tag{7.2}$$

where h is the piezometric head, p is the pressure in the soil water, γ is the weight density of the water, S is soil suction and z is the elevation above a fixed horizontal datum.

Since we are interested in this discussion only in the simpler forms of the groundwater equations, we will immediately reduce Darcy's law to its one-dimensional form. The assumption is commonly made in groundwater hydraulics that all the streamlines are approximately horizontal and the velocity is uniform with depth so that we can adopt a one-dimensional method of analysis. This is known as the *Dupuit–Forcheimer* assumption and it gives the one-dimensional form, of the equation (7.1)

Dupuit–Forcheimer assumption

$$V(x, t) = -K \frac{\partial}{\partial x} \left[h(x, t) \right] \tag{7.3}$$

where K is the hydraulic conductivity as before and h is the piezometric head. The above assumption leads immediately to the following relationship between the flow per unit width and the height of the water table over a horizontal impervious bottom as:

$$q = -Kh \frac{\partial h}{\partial x} \tag{7.4}$$

where h is the height of the water table over the impervious layer.

In order to solve any particular problem in *horizontal groundwater flow* it is necessary to combine the above equation with an equation of continuity. The one-dimensional form of the *equation of continuity* for horizontal flow through a saturated soil is

Equation of continuity

$$\frac{\partial q}{\partial x} + f \frac{\partial h}{\partial t} = r(x, t) \tag{7.5}$$

where q is the horizontal flow per unit width, h is the height of the water table i.e. the upper surface of the groundwater reservoir, f is the drainable pore space (which is initially assumed to be constant), and $r(x, t)$ is the rate of recharge at the water table. Substitution from equation (7.4) into equation (7.5) and rearrangement of the terms gives the basic equation for unsteady one-dimensional horizontal flow in a saturated soil as

$$K \frac{\partial}{\partial x} \left(h \frac{\partial h}{\partial x} \right) + r(x, t) = f \frac{\partial h}{\partial t} \tag{7.6}$$

Boussinesq equation

which is frequently referred to as the *Boussinesq equation*. The solution of this equation for both steady and unsteady flow conditions will be discussed below.

Flow through an unsaturated porous medium may also be assumed to follow Darcy's law but in this case the *unsaturated hydraulic conductivity* (K) is a function of the moisture content. In the unsaturated soil above the water table the pressure in the soil water will be less than atmospheric and will be in equilibrium with the soil air only because of the curvature of the soil water–air interface. In order to avoid continual use of negative pressures, it is convenient and is customary in discussing unsaturated flow in porous media to use the negative of the pressure head and to describe this as *soil suction* (S) or some such term. In our simplified approach we will deal only with vertical movement in the unsaturated zone and accordingly the general three-dimensional form of Darcy's law given by equation (7.1) will reduce to

Soil suction

$$V(z,t) = -K\frac{\partial}{\partial z}(-S + z) = K\frac{\partial S}{\partial z} - K \tag{7.7}$$

where both the rate of flow per unit area (V) and the vertical co-ordinate (z) are taken vertically upwards and both the unsaturated hydraulic conductivity (K) and the soil suction (S) are functions of the moisture content.

If the soil suction (S) is assumed to be a single-valued function of the moisture content (c), we can define the *hydraulic diffusivity* of the soil as

Hydraulic diffusivity

$$D(c) = -K(c)\frac{dS}{dc} \tag{7.8}$$

and rewrite equation (7.7) in the form

$$V(z,t) = -D(c)\frac{\partial c}{\partial z} - K(c) \tag{7.9}$$

which is the one-dimensional form of Darcy's law for vertical flow in an unsaturated porous medium. This formulation has the advantage that the flow equation can be written in terms of the gradient of moisture content and has the further advantage that over a given range of moisture content the variation in the hydraulic diffusivity (D) would be less than the variation in the hydraulic conductivity (K).

For unsteady vertical flow in an unsaturated soil we have as the equation of continuity:

$$\frac{\partial V}{\partial z} + \frac{\partial c}{\partial t} = 0 \tag{7.10}$$

where V is the rate of upward flow per unit area and c is the moisture content expressed as a proportion of the total volume.

A combinations of equations (7.9) and (7.10) gives us the following relationship

$$\frac{\partial}{\partial z}\left[D(c)\frac{\partial c}{\partial z}\right] + \frac{\partial}{\partial z}[K(c)] = \frac{\partial c}{\partial t} \tag{7.11}$$

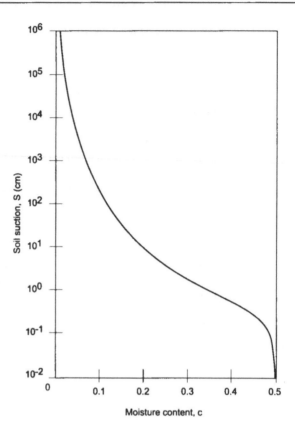

Figure 7.1. Variation of soil moisture suction (Yolo light clay — after Moore, 1939).

as the general equation for unsteady vertical flow in an unsaturated porous medium in its diffusivity form (Richards 1931). This equation will also be discussed below for both steady and unsteady flow conditions of interest in hydrologic analysis.

The solution of equation (7.11) for any particular case of unsaturated flow is far from easy due to the complicated relationship between the soil moisture suction (S) and the moisture content (c) and the complicated relationships of the unsaturated hydraulic conductivity (K) and the hydraulic diffusivity (D) with the moisture content (c). Figure 7.1 shows the variation of soil moisture suction with moisture content for a soil commonly used as an example in the literature (Moore, 1939; Constantz, 1987).

Soil moisture suction being a negative pressure head is most conveniently expressed in terms of a unit of length but is sometimes shown in the equivalent form of multiples of atmospheric pressure or as energy per unit weight. The classical form of plotting a *soil moisture characteristic* curve is in terms of the *pF* (or logarithm of the soil suction in centimetres) versus the moisture content. Figure 7.2 shows a typical relationship between hydraulic conductivity and moisture content and Figure 7.3 the

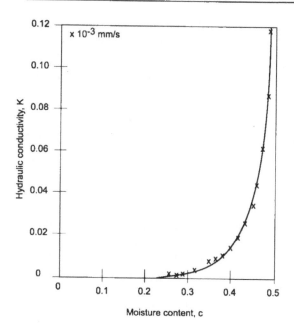

Figure 7.2. Variation of hydraulic conductivity (Yolo light clay   after Moore, 1939).

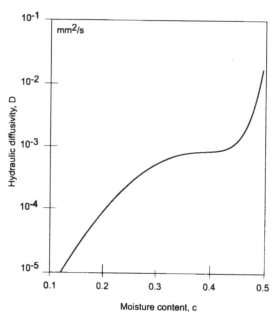

Figure 7.3. Variation of hydraulic diffusivity (Yolo light clay — after Moore, 1939).

relationship between hydraulic diffusivity and moisture content for the same soil. If the soil moisture characteristics are given empirically as in Figures 7.1 to 7.3, then the only correct approach to the solution of equation (7.11) is through numerical methods. A number of authors have suggested empirical relationships between the unsaturated hydraulic conductivity (K) or the hydraulic diffusivity (D) on the one hand and either the moisture content (c) or the soil moisture suction (S) on the other. In the case of some of these relationships, their form facilitates the solution of equation (7.11).

The simplest special case is given if we assume that both the hydraulic conductivity (K) and the hydraulic diffusivity (D) are independent of the moisture content so that equation (7.11) can be written in the special form

$$D\frac{\partial^2 c}{\partial z^2} = \frac{\partial c}{\partial t} \tag{7.12}$$

Diffusion equation

Constant D and K

which is the classical *linear diffusion equation* of mathematical physics. Solutions based on these highly simplified assumptions will be dealt with later on in this chapter, but for the moment, we are concerned with the implication of assuming both K and D to be constant. If these parameters are taken as constant in equation (7.8), which defines hydraulic diffusivity, we can integrate the latter equation and use the condition that soil moisture suction will be zero at saturation moisture content to obtain

$$S = \frac{D}{K}(c_{sat} - c) \tag{7.13}$$

which indicates that the assumption of constant values for D and K necessarily implies a linear relationship between soil section and moisture content. For our purpose the question is not so much whether the above three assumptions are accurate, but whether their use in the solution of problems of hydrologic significance gives rise to errors of an unacceptable magnitude.

A slightly less restrictive linearisation of equation (7.11) can be obtained by taking the hydraulic conductivity (K) as a linear function of moisture content (c) instead of as a constant while still retaining the hydraulic diffusivity (D) as a constant (Philip, 1968). This gives us

$$K = a(c - c_0) \tag{7.14}$$

Constant D, linear K

where c_0 is the moisture content at which conductivity is zero. For the assumptions that D is constant and K is a linear function of c, equation (7.11) becomes

$$D\frac{\partial^2 c}{\partial z^2} + a\frac{\partial c}{\partial z} = \frac{\partial c}{\partial t} \tag{7.15}$$

which is a linear convective-diffusion equation. Again the above pair of assumptions implies a particular relationship between soil moisture suction (S) and moisture content (c). The relationship is obtained by substituting a constant value of D and the value of K given by equation (7.14)

Table 7.1. Richards equation.

Model A	Model B	General case
$K = \text{constant}$	$K = a\,(c - c_0)$	$K = K(c)$
$D = \text{constant}$	$D = \text{constant}$	$D = D(c)$
$S = \dfrac{D}{K}(c_{sat} - c)$	$S = \left[\dfrac{D}{a}\right] \log_e \left[\dfrac{c_{sat} - c_0}{c - c_0}\right]$	$S = \displaystyle\int_c^{c_{sat}} \dfrac{V}{K}\, dc$
$D\dfrac{\partial^2 c}{\partial z^2} = \dfrac{\partial c}{\partial t}$	$D\dfrac{\partial^2 c}{\partial z^2} + a\dfrac{\partial c}{\partial z} = \dfrac{\partial c}{\partial t}$	$\dfrac{\partial}{\partial z}\left[D(c)\dfrac{\partial c}{\partial z}\right] + \dfrac{\partial}{\partial z}\left[K(c)\right] = \dfrac{\partial c}{\partial t}$

in equation (7.8) and integrating as before. In this case the relationship is found to be

$$S = \frac{D}{a} \log_e \left[\frac{c_{sat} - c_0}{c - c_0}\right] \tag{7.16}$$

where c_0 could be considered physically as representing the ineffective porosity, or else considered merely as a parameter chosen to give the best fit in any particular problem. The linearisation leading to equation (7.15) was used by Philip (1968) and solved for the case of ponded infiltration.

General case

The above cases can be summarised in Table 7.1. Although the third column is headed "general case", it must be remembered that the equations are all expressed in diffusivity form, which assumes that S is a single-valued function of c i.e. that there is no *hysteresis* between the wetting and the drying curves.

The subject of unsaturated flow in porous media is a wide one and the literature on it is vast. Good introductions to aspects relevant to systems hydrology are given in such publications as Domenico (1972), Corey (1977), Nielsen (1977), and De Laat (1980).

7.2 STEADY PERCOLATION AND STEADY CAPILLARY RISE

Since we are attempting a simplified analysis of the flow through the subsurface system as a whole, we will deal first with the problem of the *unsaturated zone*, the outflow from which, constitutes the inflow into the groundwater sub-system. The condition when there is no movement of

No movement of soil moisture

soil moisture in the unsaturated zone is easily seen from the examination of equation (7.17a and b) below. There will be no vertical motion at any level in the soil profile if the hydraulic potential is the same at all levels i.e. if

$$\phi(z) = -S(z) + z = \text{constant} \tag{7.17a}$$

in which $S(z)$ is the soil moisture suction at a level z above the datum. The above equation can be rearranged in a more convenient form

$$S(z) = z - z_0 \qquad (7.17b)$$

where z_0 is the elevation of the water table where the suction is by definition zero. Equation (7.17) indicates that, for the equilibrium condition of no flow at any level in the profile, the soil water suction must at every point be equal to the elevation above the water table. Consequently, at each level the moisture content must adjust itself in accordance with the soil moisture relationship (such as shown in Figure 7.1) in order to maintain this equilibrium. Thus, where no vertical movement occurs, the soil moisture profile relating moisture content to elevation will have the same shape as the curve shown in Figure 7.1.

In the case of the simplified model based on constant hydraulic conductivity (K) and constant hydraulic diffusivity (D) the variation of moisture content with level can be found from the combination of equations (7.13) and (7.17) to be

$$\frac{c}{c_{sat}} = 1 - \frac{K}{Dc_{sat}}(z - z_0) \qquad (7.18)$$

The variation of moisture content is therefore a linear one with the moisture content decreasing linearly with height above the water table. It is clear that the moisture content will reduce to zero at the height above the water table given by

$$(z - z_0) = \frac{Dc_{sat}}{K} \qquad (7.19)$$

and will have to be assumed as zero at all points above this level.

For the second special case, where the hydraulic diffusivity is taken as constant and the hydraulic conductivity is proportional to the moisture content, the variation of moisture content above the water table is given through a combination of equations (7.16) and (7.17) as

$$\frac{c - c_0}{c_{sat} - c_0} = \exp\left[\frac{-K_{sat}}{D(c_{sat} - c_0)}(z - z_0)\right] \qquad (7.20)$$

so that the moisture content decreases exponentially with level above the water table and thus only approaches a value of c_0 asymptotically.

Suppose the rain continues for a very long period of time at a constant rate that is less than the saturated hydraulic conductivity of the soil — an unlikely event. We would get a condition of *steady percolation* to the water table with the rate of *infiltration* at the surface (f) equal to the rate of *recharge* (r) at the water table. For these conditions equation (7.9) would take the form

Steady percolation

$$-f = V(z) = -D(c)\frac{dc}{dz} - K(c) = -r \qquad (7.21)$$

where the derivative of moisture content with respect to elevation can be written as an ordinary rather than a partial differential, since there is no variation with respect to time. We can separate the variables in equation (7.21) to obtain:

$$dz = \frac{D(c)}{f - K(c)} dc \tag{7.22}$$

which can be integrated to give

$$z - z_0 = \int_{c(z)}^{c_{sat}} \frac{D(c)}{K(c) - f} dc \tag{7.23}$$

If the functions $K(c)$ and $D(c)$ are known, either analytically or numerically, then equation (7.23) can be integrated in order to obtain the value of the level above the water table at which any particular value of moisture content will occur.

Constant D and K

For the simplest case where the hydraulic conductivity (K) and the hydraulic diffusivity (D) are assumed to be constant, equation (7.23) immediately integrates to

$$z - z_0 = \frac{D}{K - f}(c_{sat} - c) \tag{7.24}$$

which can be rearranged to give the moisture content explicitly in terms of the elevation as

$$\frac{c}{c_{sat}} = 1 - \left(\frac{K - f}{Dc_{sat}}\right)(z - z_0) \tag{7.25}$$

which is the solution of equation (7.21) for steady *downward percolation* in a soil with constant K and D. Thus in this special case, the moisture content distribution at a steady rate of percolation is still linear with the height above the water table, but with a slope proportional to the difference between the hydraulic conductivity and the steady percolation rate (which is equal to the rate of infiltration at the surface and the rate of recharge at the water table).

Constant D, linear K

For the second type of linearisation where the hydraulic conductivity (K) is taken as proportional to the moisture content and the hydraulic diffusivity is taken as a constant, equation (7.23) will integrate to

$$(z - z_0) = \frac{D(c_{sat} - c_0)}{K_{sat}} \log_e \left[\frac{K_{sat} - f}{K - f}\right] \tag{7.26}$$

where f is the steady infiltration rate and the other symbols are as in equation (7.16). The above equation can be rearranged to give the moisture content in terms of the elevation as

$$\frac{c - c_0}{c_{sat} - c_0} = \frac{f}{K_{sat}} + \left[1 - \frac{f}{K_{sat}}\right] \exp\left[-\frac{K_{sat}}{D(c_{sat} - c_0)}(z - z_0)\right] \tag{7.27}$$

which is again seen to be exponential in form. This time for a very deep water table the moisture content is asymptotic to the value c where $(c - c_0)$ is the same proportion of the saturation moisture content $(c_{sat} - c_0)$ as the percolation is of the saturated hydraulic conductivity.

After the rainfall has ceased, the water in the unsaturated will be depleted by evaporation at the ground surface. For long continuous periods without precipitation, it is possible that an equilibrium condition of *Capillary rise* *capillary rise* from the groundwater to the surface could develop in the case of shallow water tables. For true equilibrium, the rate of supply of water at the water table would have to be equal to the upward transport of water at any level and to the evaporation rate (e) at the surface. For such a condition, equation (7.9) would take the form

$$V(z, t) = e = -D(c)\frac{\partial c}{\partial z} - K(c) \tag{7.28}$$

For the case of steady upward movement of water, the water level for any given moisture content can be obtained from the integration

$$z - z_0 = \int_{c(z)}^{c_{sat}} \frac{D(c)}{K(c) + e} dc \tag{7.29}$$

which, as might be expected, is the same as equation (7.23) for the steady downward percolation except that the sign of the term representing the *Evaporation* steady rate of *evaporation* (e) is opposite to the sign for the steady infiltration (f). Consequently, the moisture content distribution with elevation *Constant D and K* for the case where both the hydraulic conductivity (K) and the hydraulic diffusivity (D) are taken as constant would be

$$\frac{c}{c_{sat}} = 1 - \frac{(K + e)}{Dc_{sat}}(z - z_0) \tag{7.30}$$

which is also a linear variation of moisture content with height but with a steeper gradient, which would be expected as the gradient of soil moisture suction has to act against gravity in this instance. A similar situation arises for the second linear model. In this case, the hydraulic conductivity (K) *Constant D, linear K* is taken as a linear function of the moisture content (c), and the variation of moisture content with elevation can be obtained by substituting for the steady infiltration rate (f) in equation (7.27) the steady rate of evaporation (e) with the sign reversed. This gives us

$$\left[\frac{c - c_0}{c_{sat} - c_0}\right] = \left[1 + \frac{e}{K_{sat}}\right]\exp\left[-\frac{K_{sat}}{D(c_{sat} - c_0)}(z - z_0)\right] - \frac{e}{K_{sat}} \tag{7.31}$$

which is again exponential in form.

It is clear from the form of equation (7.29), that for high rates of evaporation (e), the calculated value for the elevation above the water table, corresponding to a vanishingly small moisture content, might be considerably less than the elevation of the surface of the column of unsaturated soil. This suggests that there might be a *limiting rate of evaporation* above which the capillary rise would be unable to supply sufficient water, because the soil would become completely dry and unable to transfer water upwards to the surface. Gardner (1958) showed that if the unsaturated hydraulic conductivity is taken as a function of the soil moisture of the form

$$K = \frac{a}{b + S^n} \tag{7.32}$$

Limiting rate of evaporation

then, for any given value of the exponent n, the limiting rate of evaporation would be given by equation

$$\frac{e_{limiting}}{K_{sat}} = \frac{constant}{(z_s - z_0)^n} \tag{7.33}$$

where n has the same exponent as in equation (7.32), z_s is the elevation of the surface, z_0 the elevation of the water table, and the constant depends only on the value of n. Accordingly, for the case studied by Gardner, the limiting rate of evaporation is inversely proportional to the appropriate power of the depth of the water table.

This concept of limiting evaporation rate can be applied to the linear models, on which we are concentrating in this discussion, even though they are not special cases of equation (7.32). Thus, an examination of equation (7.30), which applies to the highly simplified model based on

Constant D and K

constant values of hydraulic conductivity (K) and hydraulic diffusivity (D), reveals that the value of the moisture content will be zero for a surface elevation of z_s if the evaporation reaches the limiting value of

$$\frac{e_{limiting}}{K} = \frac{Dc_{sat}}{K(z_s - z_0)} - 1 \tag{7.34}$$

Constant D, linear K

For a high limiting evaporation rate, this rate is approximately inversely proportional to the depth from the surface to the water table. For the case where the hydraulic conductivity is taken as proportional to the moisture content, we can deduce from equation (7.31) that the *limiting rate of evaporation* is given by:

$$\frac{e_{limiting}}{K_{sat}} = \frac{1}{1 - \exp\left[\frac{K_{sat}}{D(c_{sat} - c_0)}\right](z_s - z_0)} \tag{7.35}$$

7.3 FORMULAE FOR PONDED INFILTRATION

The classical problem in the unsteady vertical flow in the unsaturated zone is that of *ponded infiltration*. In this case, the surface of the soil column is assumed to be saturated, so that the rate of infiltration is soil-controlled and independent of the rate of precipitation. The basic equation (7.11)

Table 7.2. Solutions for concentration boundary condition.

	Constant K	Linear K	Non-linear K
Delta-function D		Green and Ampt (1911)	
Constant D	Carslaw and Jaeger (1946)	Philip (1969)	Philip (1974)
Fujita D	Fujita (1952)	Not solved	Not solved

can be transformed from an equation in $c(z, t)$ to an equation in a single transformed variable $c(z^2/t)$. To obtain a solution in this transformed space, it is necessary to reduce the two boundary conditions $c(0, t) = c_{sat}$ and $c(1, t) = c_1$ and the single initial condition $c(z, 0) = c_0$ to two boundary conditions in the new variable $c(z^2/t)$. This is possible for the case of an infinite column with a constant initial moisture condition $c(z^2/t) = c_0$ and consequently analytical solutions can be sought for these conditions.

On the basis of the above transformation a number of such analytical solutions can be derived both for the case of ponded infiltration and for the case of constant precipitation under pre-ponding conditions. The latter solutions for the pre-ponding case give results for the time to surface saturation (and subsequent ponding) and for the distribution of moisture content with depth at this time. The special cases in Table 7.1 can be expanded to cover these known solutions for both ponded infiltration (Table 7.2) which is *soil-controlled*, and for pre-ponding infiltration (Table 7.3) which is *atmosphere-controlled* (Kühnel et al., 1990a, b).

Pre-ponding infiltration

It can be demonstrated in all cases of initial pre-ponding constant inflow that the shape of the *moisture profile at ponding* is closely approximated by the shape for the same total moisture in the column under ponded conditions. (Kühnel 1989; Kühnel et al., 1990a, b; Dooge and Wang, 1993). This is illustrated in Figure 7.4 for the special cases shown in Tables 7.2 and 7.3.

In practice, the soil moisture rarely attains an equilibrium profile of the type discussed in the previous section. Conditions of constant rainfall, or of constant evaporation, do not persist for a sufficient period for such an equilibrium situation to develop. With alternating precipitation and evaporation, there will be continuous changes in the soil moisture profile, and unsteady movement of water either upwards or downwards in the soil. A distinct possibility arises of a combination of upward movement near the surface under the influence of evaporation and simultaneous downward percolation in the lower layers of the soil.

A major point in applied hydrology is the rate at which infiltration will occur during surface runoff i.e. in the question of the extent to which the total precipitation should be reduced to effective precipitation in attempting to predict direct storm runoff. It is important to distinguish between the infiltration capacity of the soil at any particular time and the actual

Infiltration capacity

infiltration occurring at that time. Infiltration capacity is the maximum

Table 7.3. Solutions for flux boundary conditions.

	Constant K	Linear K	Non-linear K
Delta-function D		Mein and Larson (1973)	
Constant D	Braester (1973)	Braester (1973)	Clothier et al. (1981)
Fujita D	Knight and Philip (1974)	Rogers et al. (1983)	Sander et al. (1987)

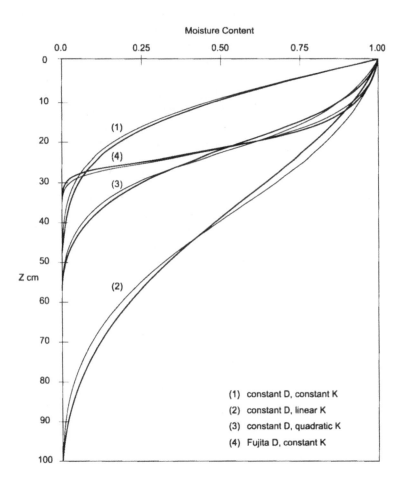

Figure 7.4. Comparison of profiles at ponding.

rate at which the soil in a given condition can absorb water at the surface. If the rate of rainfall or the rate of snow melt is less than the infiltration capacity, the actual infiltration will be equal to the actual rate of rainfall or of snow melt, since the amount of moisture entering the soil cannot exceed the amount available.

A number of empirical formulae for infiltration capacity have been proposed from time to time. Kostiakov (1932) proposed the following formula for the initial high rate of infiltration into an unsaturated soil

$$f = \max \left[\frac{a}{t^b}, K_{sat} \right] \qquad (7.36)$$

where f is the rate of infiltration up to the time when the infiltration rate becomes equal to the saturated permeability of the soil, t is the time elapsed since the start of infiltration and a and b are empirical parameters. It will be seen later that many of the simpler theoretical approaches to the problem of ponded infiltration give solutions which indicate that the initial high rate of infiltration follows the Kostiakov formula with the value of b equal to 1/2. Other values of b have been used and the Stanford Watershed Model uses a value of $b = 2/3$.

Horton (1940) suggested, on the basis of certain physical arguments, that the decrease in infiltration capacity with time should be of exponential form and suggested the formula

$$f - f_c = (f_0 - f_c) \exp(-kt) \qquad (7.37)$$

Excess infiltration

where f is the rate of infiltration capacity, f_c is the ultimate rate of infiltration capacity, f_0 is the initial rate of infiltration capacity and k is an empirical constant. Holtan (1961) suggested that the rate of excess infiltration (i.e., the rate of infiltration capacity minus the ultimate rate of infiltration capacity) in the early part of a storm could be related to the volume of potential infiltration F_p by an equation of the form

$$f - f_c = a(F_p)^n \qquad (7.38)$$

where a and n are empirical constants. Overton (1964) showed that if we take $n = 2$ in equation (7.38), the rate of infiltration capacity can be expressed explicitly as a function of time in the following form

$$f = f_c \sec^2 \left[\sqrt{af_c}(t_c - t) \right] \qquad (7.39)$$

where t_c is the time taken for the infiltration capacity rate to fall to its final value f_c and is given by:

$$t_c = \frac{1}{\sqrt{af_c}} \tan^{-1} \left[F_c \sqrt{\frac{a}{f_c}} \right] \qquad (7.40)$$

where F_c is the ultimate volume of infiltration, which is the same as the initial volume of potential infiltration.

We now turn from *phenomenological models involving* empirical formulae based on analysis of field observations to models involving theoretical formulae based on the principles of soil physics and hence on the equations described in Section 7.1 above. We saw in that section that the unsteady movement of moisture in a vertical direction in the unsaturated zone of the soil is governed by equation (7.11) which is repeated here

$$\frac{\partial}{\partial z}\left[D(c)\frac{\partial c}{\partial z}\right] + \frac{\partial}{\partial z}\left[K(c)\right] = \frac{\partial c}{\partial t} \tag{7.11}$$

Ponded infiltration

If we take the case of heavy rainfall following a relatively dry period, we will be concerned with the problem of ponded infiltration. This problem can be formulated in terms of the above equation and an appropriate set of boundary conditions. If the surface is saturated throughout the period of concern, we have the boundary condition at the surface:

$$c(z_s, t) = c_{sat} \quad \text{for all } t \tag{7.41}$$

where z_s is the elevation of the surface. Since the soil is, by definition, saturated at the water table we get the boundary condition at the water table as:

$$c(z_0, t) = c_{sat} \quad \text{for all } t \tag{7.42}$$

where z_0 is the elevation of the water table. The initial condition will be given by

$$c(z, 0) = c_1(z) \tag{7.43}$$

where $c_1(z)$ is the initial distribution of soil moisture content in the unsaturated zone. The problem as posed above is far from easy to solve, since equation (7.11) is non-linear and the functions $D(c)$ and $K(c)$ may be only known empirically, or may require complicated expressions for their representation. Accordingly, comprehensive discussion of the solution of the problem of ponded infiltration (Philip, 1969) is well outside the scope of the present chapter. However some simplified approaches are discussed below.

If we start with the simplest form of equation (7.11), i.e. that obtained by assuming both D (the hydraulic diffusivity), and K (the hydraulic conductivity) to be constant, we obtain the linear diffusion equation already given above as equation (7.12) and repeated here:

$$D\frac{\partial^2 c}{\partial z^2} = \frac{\partial c}{\partial t} \tag{7.12}$$

What is required is a solution of this equation for $c(z, t)$ which will satisfy the boundary conditions given by equations (7.41) and (7.42) and the initial condition given by equation (7.43).

Actually it is more convenient to solve the equation in terms of the depth below the soil surface x rather than in terms of the elevation above a fixed datum z, i.e. to make the transformation

$$x = z_s - z \tag{7.44}$$

This transformation results in the basic differential equation

$$D\frac{\partial^2 c}{\partial x^2} = \frac{\partial c}{\partial t} \tag{7.45}$$

which is seen to be exactly the same form as equation (7.12). The boundary condition at the surface given by equation (7.41) becomes

$$c(0, t) = c_{sat} \tag{7.46}$$

and the boundary condition at the water table becomes

$$c(x_0, t) = c_{sat} \tag{7.47}$$

where x_0 is the depth of the water table below the soil surface. The initial condition is now written as

$$c(x, 0) = c_0(x) \tag{7.48}$$

Boltzman transformation

Equation (7.45) can be converted to an ordinary differential equation by means of the Boltzman transformation, which we write as

$$n = xt^{-1/2} \tag{7.49a}$$

which converts equation (7.45) above to

$$D\frac{d^2 c}{dn^2} + \frac{n}{2}\frac{dc}{dn} = 0 \tag{7.49b}$$

which is a non-linear ordinary differential equation rather than a linear partial differential equation.

But the complete problem can only be solved in this way, if the three conditions represented by equations (7.46), (7.47) and (7.48) can be reduced to two conditions in terms of the transformed variable n. The boundary condition for the surface clearly transforms to the condition

$$c(n) = c_{sat} \quad \text{for } n = 0 \tag{7.50}$$

The other two conditions represented by equations (7.47) and (7.48) can obviously be reduced to a single condition if we take the initial soil moisture content distribution as uniform and assume the depth to the water table x_0 to be infinitely large. For these two assumptions we have the second boundary condition as

$$c(n) = c_1 \quad \text{for } n = \infty \tag{7.51}$$

which imposes the constant moisture content c_1 at $x = \infty$ (and therefore at $n = \infty$) for all values of t and also sets the moisture content equal to the constant value c_1 for $t = 0$ (and consequently $n = \infty$) for all values of x. The assumption of a constant moisture content at all depths below the surface as the initial condition, can be inferred from equation (7.7) in Section 7.1 above, if the initial downward percolation is occurring at a rate equal to the hydraulic conductivity corresponding to the initial moisture content.

For the special assumptions listed above, the linear partial differential equation given by equation (7.45) can be solved for the boundary conditions given by equation (7.49), (7.50) and (7.51) to give the value of the moisture content in terms of the transformed variable n (Childs, 1936). The total amount of infiltration after a given time t can be calculated from the increase in moisture content in the infinite soil column i.e.

$$F = \int_{c_1}^{c_{sat}} x \, dc + f_1 t \tag{7.52}$$

where x is the given level below the surface and f_1 is the initial rate of infiltration which gives rise to the initial constant moisture content c_1. Since the solution of equation (7.45) gives the moisture content in terms of the transformed variable n, x will be given as the product of the square root of t multiplied by a function of the moisture content at that level. Insertion of the solution for x in equation (7.52) and integrating gives the total infiltration F as another function of the initial moisture content multiplied by the square root of the elapsed time. It can be shown for *Constant D and K*   constant D and K, that the solution for total infiltration is given by

$$F = (c_{sat} - c_1)\sqrt{\frac{4Dt}{\pi}} + f_1 t \tag{7.53}$$

where D is the hydraulic diffusivity (assumed to be constant) and f_1 is the initial infiltration, which is equal to the hydraulic conductivity K_1 corresponding to the initial moisture content c_1. The rate of infiltration for the ponded condition can be obtained by differentiating equation (7.53) to obtain

$$f = (c_{sat} - c_0)\sqrt{\frac{D}{\pi t}} + f_1 \tag{7.54a}$$

which suggests that the initial high rate of infiltration varies inversely with the square root of the elapsed time. The form of equation (7.45) assumes that the hydraulic conductivity is a constant and that the hydraulic diffusivity is also constant. We saw in Section 7.1 that these two assumptions imply the following expression for the relationship between soil suction and moisture content

$$S = \frac{D}{K}(c_{sat} - c) \tag{7.13}$$

Accordingly we can express this initial high rate infiltration capacity, which is given by equation (7.54a), in terms of initial moisture content c_1 and the hydraulic conductivity K_1 as follows

$$f = \sqrt{\frac{K_1(c_{sat} - c_1)S_1}{\pi t}} + f_1 \tag{7.54b}$$

where S_1 is the soil moisture suction corresponding to the initial moisture content c_1. Alternatively we could express it in terms of hydraulic conductivity and hydraulic diffusivity as

$$f = \frac{K_1 S_1}{\sqrt{\pi D t}} + f_1 \tag{7.54c}$$

While the forms given by equations (7.54b) and (7.54c) above are useful for comparative purposes, the original form of equation (7.54a) is the most useful in practice. It indicates clearly that the infiltration capacity is initially infinite and decreases inversely as the square root of the elapsed time and ultimately reaches a constant value equal to the hydraulic conductivity at the initial percolation rate. The dependence of the rate of infiltration on the initial soil conditions appears as a direct proportionality between the rate of infiltration and the moisture deficit $(c_{sat} - c_1)$.

Constant D, linear K   If instead of assuming the hydraulic conductivity to be constant, we take it as a linear function of the moisture content, the equation obtained is the linear convective diffusion equation as indicated by equation (7.15) in Section 7.1

$$D\frac{\partial^2 c}{\partial z^2} + a\frac{\partial c}{\partial z} = \frac{\partial c}{\partial t} \tag{7.15}$$

where D is the constant hydraulic diffusivity and a is the coefficient of the moisture content in the equation for the hydraulic conductivity given by equation (7.14). The above equation was solved for the boundary conditions of saturation at the surface, an infinite depth to the water table and a constant initial moisture content at all depths below the surface by Philip (1968). The solution is necessarily more complex and the rate of infiltration is found to be

$$f - K_{sat} = \frac{K_{sat} - K_1}{2}\left[\frac{\exp\left[-\frac{a^2 t}{4D}\right]}{\sqrt{\pi \frac{a^2 t}{4D}}} - \text{erfc}\left[\sqrt{\frac{a^2 t}{4D}}\right]\right] \tag{7.55}$$

where erfc is the complementary error function.

For small values of t the solution given by equation (7.55) above can be expanded as a power series in $t^{1/2}$ to give:

$$f - K_{sat} = \frac{K_{sat} - K_1}{2\sqrt{\pi}}\left[\sqrt{\frac{4D}{a^2 t}} - \sqrt{\pi} + \sqrt{\frac{a^2 t}{4D}} \cdots\right] \tag{7.56}$$

If the value of t is very small, then we probably obtain a good approximation by using only the first term inside square brackets in the above series. If only the first term is taken, the resulting expression is identically equal to that given by equation (7.54a) above. For slightly longer times it might be necessary to include a second term in the series and in this case the equation (7.56) would be approximated by

$$f = (c_{sat} - c_1)\sqrt{\frac{D}{\pi t}} + \frac{K_1 + K_{sat}}{2} \tag{7.57}$$

so that the only modification is in the constant term. For large values of t, it can be shown (Philip, 1968) that the general solution given by equation (7.55) is approximated closely by

$$f = (c_{sat} - c_1)\left(\frac{D}{\pi t}\right)^{3/2} \exp\left[-\frac{a^2 t}{4D}\right] + K_{sat} \tag{7.58}$$

For very large values of t, the exponential term in the first term on the right hand side of equation (7.58) will approach zero and give as the ultimate value of the infiltration rate, the saturated permeability K_{sat}.

In 1911, Green and Ampt proposed a formula for infiltration into the soil based on an analogy of uniform parallel capillary tubes. In fact, the treatment of the problem along the lines suggested by them is not dependent on this specific analogy. As pointed out by Philip (1954), it requires only the assumption that the wetting front which travels down to the soil, may be taken as a sharp discontinuity, which separates an upper zone of constant higher moisture content c_2 from the original dry soil of constant initial moisture content c_1. The rate of percolation for the upper part of the soil i.e. the wetted part may be written as

Wetting front

$$V(x, t) = K_2\left[\frac{\phi_2 - \phi_1}{x}\right] \tag{7.59}$$

where x is the depth of penetration of this *wetting front* K_2 is the hydraulic conductivity at the moisture content of the upper zone, and ϕ_2 and ϕ_1 are the values of the hydraulic potential in the upper (wetted) zone and the lower (unwetted) zone respectively. The hydraulic potential at the top of the column relative to the surface is given as

$$\phi_2 = H \tag{7.60}$$

where H is the depth of ponding on the surface. The hydraulic potential (relative to the surface) immediately below the discontinuous wetting front will be equal to

$$\phi = \frac{P_1}{\gamma} + z_1 = -S_1 - x \tag{7.61}$$

where S_1 is the suction ahead of the wetting front, which for a dry soil may be taken as the suction at air entry potential.

Substituting from equations (7.60) and (7.61) into equation (7.59) we obtain

$$V(x, t) = K_2\left[\frac{H + S_1 + x}{x}\right] \tag{7.62}$$

Wetted zone

for the percolation rate in the upper or wetted zone, which must be the same at all levels within this zone if the moisture content is constant within the zone. Since the upper wetted part of the soil is assumed to have a constant mean moisture content (c_2) and the lower unwetted part to have a constant mean moisture content c_1, we can write an equation of continuity for the wetted zone as

$$f(t) = (c_2 - c_1)\frac{dx}{dt} + f_1 \tag{7.63}$$

which connects the infiltration at the surface, the rate of downward travel of the wetting front and the rate of initial infiltration f_1, which must be equal to K_1 for c_1 to be constant. Since the rate of infiltration given by equation (7.63) is equal to the rate of percolation in the wetted zone given by equation (7.62) we can combine the two equations to write

$$(c_2 - c_1)\frac{dx}{dt} = K_2\left[\frac{H + S_a}{x}\right] + K_2 - K_1 \tag{7.64}$$

The above equation can be integrated to give

$$t = \left[\frac{c_2 - c_1}{K_2 - K_1}\right]x - \frac{K_2(H + S_a)}{K_2 - K_1}\log_e\left[1 + \frac{(K_2 - K_1)}{K_2(H + S_a)}x\right] \tag{7.65a}$$

Green–Ampt

which is the *Green–Ampt solution* for constant initial moisture content.

Equation (7.65a) has the disadvantage that it relates the depth of penetration x to the time elapsed t in implicit form and so makes it difficult to obtain the rate of infiltration from equation (7.63) as an explicit function of time. However, the infiltration rate for small values of t and for large values of t can be deduced. For very large values of t the depth of penetration x will become larger and larger compared to the other terms in the numerator of (7.62), i.e. $(H + S_a)$ and accordingly the rate of downward percolation and of infiltration at the surface will approach the constant value K_2.

The behaviour of the solution for small values of t can be seen most conveniently by rearranging equation (7.65a) in dimensionless form and expanding the second term on the right hand side as an infinite series. This converts equation (7.65a) to the form

$$\frac{(K_2 - K_1)^2}{K_2(H + S_a)(c_2 - c_1)}t = \sum_{r=2}^{\infty}\frac{(-1)^r}{r}\left[\frac{(K_2 - K_1)}{K_2(H + S_a)}x\right]^r \tag{7.65b}$$

Small values of t

It is clear that for small values of t, and consequently for small values of x, that the series on the right hand side of equation (7.65b) will converge rapidly. If t is sufficiently small so that only the first term (i.e. the term for $r = 2$) needs to be considered, we will have, after cancelling common factors on the two sides of the equation,

$$x = \sqrt{\frac{2K_2(H + S_a)}{(c_2 - c_1)}} t \tag{7.66}$$

Substitution from equation (7.66) into equation (7.63) gives us the infiltration as an explicit function of time in the form

$$f(t) = \sqrt{\frac{K_2 (H + S_a)(c_2 - c_1)}{2t}} \tag{7.67}$$

It is clear from equation (7.65b) that if the difference between the hydraulic conductivity of the wetted soil K_2 and the hydraulic conductivity of the unwetted soil K_1 becomes vanishingly small, all the terms in the series for $r > 2$ will become negligible. Consequently for this case, the infiltration rate at all times will be given by equation (7.67) above. It will be noted that equation (7.67) derived from the Green–Ampt approach gives a result which only differs from equation (7.54b) (which was based on the assumption of constant hydraulic conductivity and constant hydraulic diffusivity) in regard to the numeric value which appears in the denominator.

Philip

A more complete theory of ponded infiltration allowing for the concentration-dependent diffusivity and for the gravity term has been developed by Philip (Philip 1957a, Philip 1957b). Philip showed that the equation relating the depth of penetration of a given moisture content with time can be represented by a series of the form.

$$x(c, t) = \sum_{m=1}^{\infty} a_m(c) t^{m/2} \tag{7.68}$$

which states that, for the range of t and values of hydraulic conductivity and hydraulic diffusivity of interest to soil scientists, the above series converges so rapidly that only a few terms are required for an accurate solution. More recently, Salvucci (1996) has shown that the convergence can be improved if the elapsed time t in equation (7.68) is replaced by a transformed time $t' = t/(t + a)$ where the parameter a depends on the soil characteristics. The solution given above in equation (7.66) is seen to correspond to the first term of a series of the type given in equation (7.68).

The relationship represented by equation (7.52) given earlier can be used to obtain a series expression for the total infiltration F in terms of time for any given initial moisture content c_0. The resulting series converges except for very large values of the elapsed time t. Philip suggested that for the most practical purposes only the first two terms are required so that we can write

$$F = St^{1/2} + At \tag{7.69}$$

where S is a property of the soil and the initial moisture content, which Philip called sorptivity, and the second parameter A is also a function of the soil and the initial moisture content. In a series of papers, Philip (1957a,b) discussed the implications of the nature of the soil profile, the effect of surface ponding and other factors, on the solution given by this approach.

It must be emphasised that the solutions given above all relate to one particular formulation of the infiltration problem. In every case, the analysis is made on the basis of an infinitely deep soil profile (not subject to hysteresis) with uniform initial moisture content, into which infiltration takes place as a result of saturation of the surface. Such a stylised case would have to be modified in several respects before it would correspond closely to conditions of actual catchments. In practice, the above theoretical solutions would be modified by the presence of the water table at some finite depth, by the actual moisture distribution of the profile at the instant that the surface was first saturated. This would also depend on (a) the previous history of moisture distribution, (b) the movement in the profile itself, (c) distinct layers in the soil profile which might give rise to interflow, (d) on the possibility of shrinking and swelling in the soil, and so on. Nevertheless, as in many other instances in hydrology, a simple model can be explored in order to get a feel for phenomena under study, and may subsequently be used as the basis of a more complex model.

A number of comparisons have been made of the various solutions of both analytical and numerical solutions for ponded infiltration and initial high rate infiltration (e.g. Wang and Dooge, 1994). Comparisons have also been made between the moisture profiles in the soil for (a) high rate infiltration followed by ponded infiltration, and (b) ponded infiltration throughout the period of interest, making use of a *compression or con-*

densation of the time scale to match the volume infiltrated up to the time of ponding. The subsequent profiles (and consequently fluxes) are not identical but are close approximations of one another (Dooge and Wang, 1993) as shown in Figure 7.4.

7.4 SIMPLE CONCEPTUAL MODELS OF INFILTRATION

It can be shown that a number of infiltration equations derived either empirically or from simple theory can also be derived by postulating a relationship between the rate of infiltration and the volume of either actual or potential infiltration (Overton, 1964; Dooge, 1973). Apart from its intrinsic interest, the formulation of infiltration as a relationship between a rate of infiltration and a volume of actual or potential infiltration would appear to have many advantages in the formulation and computation of conceptual models of the soil moisture phase of the catchment response and its simulation.

Volume of infiltration

If we wish to relate the rate of infiltration to the volume of infiltration which has occurred, the relationship must be such that the rate of infiltration decreases with the volume of water infiltrated in order to reproduce the observed behaviour of the decrease of infiltration with time. On simple way of accomplishing this is to take the rate of infiltration as inversely proportional to some power of the volume of infiltration up to that time i.e.

$$f = \frac{a}{F^c} \tag{7.70}$$

where f is the infiltration rate at given time, F is the total volume of infiltration at the same time, and a and c are empirical constants. Taking advantage of the fact that the rate of infiltration is the derivative with respect to time of the volume of infiltration, equation (7.70) can be integrated readily to express the corresponding rate of infiltration explicitly as a function of time. The result of this integration is

$$f = \left[\frac{a^{1/c}}{(c+1)t} \right]^{c/(c+1)} \tag{7.71}$$

in which the infiltration rate is seen to have the required feature of declining with time as long as the parameter c is non-negative. Equation (7.71) derived from postulating a relationship between infiltration rate and infiltration volume is seen to have the same form as the empirical equation proposed by Kostiakov (1932). For the value of $c = 1$ this particular conceptual model would give a variation of infiltration rate which is inversely proportional to the square root of elapsed time which corresponds to a number of the simple theoretical models discussed in Section 7.3. A relationship of the type indicated by equation (7.70) is used in the Stanford Watershed Model and a value of $c = 2$ is customarily used.

Kostiakov

Final constant infiltration rate

The above simple conceptual model can easily be modified to allow for a final constant infiltration rate by relating the excess infiltration rate above this final rate to the volume of excess infiltration i.e. the total volume of such excess infiltration which has accumulated. If we modify equation (7.70) in this way we obtain a conceptual model represented by

$$f - f_c = \frac{a}{(F - f_c t)^c} \tag{7.72a}$$

in which f_c is the constant rate of infiltration. This can more conveniently be written in terms of the effective infiltration $f_e = f - f_c$ as

$$f_e = \frac{dF_e}{dt} = \frac{a}{F_e^c} \tag{7.72b}$$

which can be integrated as before to give

$$F_e = [(c+1) \, at]^{1/(c+1)} \tag{7.73a}$$

which gives the volume of excess infiltration F_e as a function of time. The latter equation can be written in terms of actual infiltration as

$$F = [(c + 1)at]^{1/(c+1)} + f_c t \tag{7.73b}$$

Philip

For the value of $c = 1$, this corresponds to the simplified equation of Philip given by equation (7.69) above and the parameters S and A in that equation can be related easily to the parameters a and c in equation (7.72).

If the rate of excess infiltration is taken as inversely proportional to the volume of total infiltration, i.e.

$$f - f_c = \frac{a}{F} \tag{7.74}$$

it can be shown that the relationship between the total volume of infiltration and time is given implicitly by

$$t = \frac{a}{f_c^2}\left[\frac{F}{a/f_c} - \log_e\left\{1 + \frac{F}{a/f_c}\right\}\right] \tag{7.75}$$

Green–Ampt

which is seen to be the same form as the Green–Ampt solution given by equation (7.65a) above.

If we relate the rate of infiltration to potential infiltration volume, the simplest relationship, which we can postulate, is

$$f = aF_p \tag{7.76a}$$

where F_p is the potential infiltration volume, i.e. the ultimate volume of infiltration minus the volume of infiltration at any particular time. The relationship can be written as

$$f = \frac{dF}{dt} = a(F_o - F) \tag{7.76b}$$

where F_o is the ultimate volume of infiltration; or in terms of the initial infiltration rate $f_0 = aF_o$ as

$$f = \frac{dF}{dt} = f_0 - aF \tag{7.76c}$$

The latter equation can be solved to give the following expression for the rate of infiltration

$$f = f_0 \exp(-at) \tag{7.77}$$

If we wish to obtain an expression involving an ultimate non-zero constant rate of infiltration (f_c), we need to relate the rate of infiltration excess to the potential volume of infiltration excess, i.e. to write equation (7.76c) in the more general form

$$f - f_c = (f_0 - f_c) - a(F - f_c t) \tag{7.78}$$

which can be integrated to give the rate of infiltration f as an explicit function of time of the following form

$$f = f_c + (f_0 - f_c)\exp(-at) \tag{7.79}$$

which is the same form as the Horton infiltration equation.

Overton

Overton (1964) proposed the relationship

$$f - f_c = aF_p^2 \tag{7.80}$$

which can be solved to give the explicit relationship of equation (7.39) already mentioned

$$f = f_c \sec^2\left[\sqrt{af_c}(t_c - t)\right] \tag{7.42}$$

where t_c is the time taken for the infiltration to fall to the ultimate constant rate f_c and is given by equation (7.40) in Section 7.3.

In Chapter 5 we made extensive use of the simple conceptual component of a linear reservoir, which is defined as an element in which the outflow is directly proportional to the storage in the reservoir. Equation (7.76) above can be considered to represent a conceptual element in which the inflow to the element is proportional to the storage deficit in the element. Hence, it might be regarded as a special conceptual element, which could fittingly be referred to as a *linear absorber*. On this basis, the relationship indicated by equation (7.78) could be considered as consisting of a linear absorber preceded by a constant rate of overflow, which diverts water at a rate f_c around the absorber and feeds at this rate into the groundwater reservoir, even when the field moisture deficit is not satisfied. By analogy, equation (7.70) might be considered as being represented by a second type of conceptual element in which the inflow into the element is inversely proportional to some power of the amount of inflow which has taken place already. This might be referred to as an inverse absorber or some similar term. Just as arrangements of linear reservoirs were useful in building conceptual models of direct storm runoff, so also simple arrangements of linear absorbers or linear inverse absorbers might be useful in modelling the subsurface flow in the unsaturated zone.

Linear absorber

Inverse absorber

An interesting conceptual model (Zhao Dihua and Dooge, 1990) of the unsaturated zone, incorporating infiltration under surface ponding and outflow is obtained by combining the single linear reservoir described by equations (5.9) to (5.14) with the linear version of the conceptual model given by equation (7.72). If $W(t)$ is the water content of the unsaturated zone, the water balance can be written as

$$\frac{dW}{dt} = \frac{a}{W} + \frac{W}{b} \tag{7.81}$$

where a is an infiltration parameter and b is an outflow parameter. Since equation (7.81) is linear in W^2 an analytical solution is available for certain cases. In general a method of soil moisture accounting can be applied. This has been done for the Gauwu experimental basin (2.5 hectares) in Zhejiang Province and compared with the measured outflow (Zhao Dihua and Dooge, 1990). The Nash–Sutcliffe efficiency was found to be 96.3%.

7.5 EFFECT OF THE WATER TABLE

If any of the above simple models are to be used as components in the simulation of the total catchment response, they must be adapted to allow for (a) the effect of the level of the water table, (b) the redistribution of moisture in the soil profile following the end of a rainstorm, and other factors.

The model, which seems to offer the best hope of taking account of the effect of the water table, is that based on the Green–Ampt approach. The solution discussed above in equations (7.59) to (7.67) applies to the case where there is a constant initial moisture content (c_1) in the soil profile. If the moisture content of the profile is constant, the soil moisture suction will also be constant. In accordance with equation (7.17) in Section 7.1, the rate of infiltration at the surface and the downward movement throughout the profile must be equal to the hydraulic conductivity at the initial moisture content K. Since the soil moisture content will be equal to the saturation value at the water table, we must either postulate a water table at infinite depth, or else a discontinuity in moisture content at the water table.

The assumption of a constant initial moisture content gives rise to the series solution of equation (7.65b) for the general case where no special soil moisture characteristics are specified. We saw in Section 7.3 that if we make the simple assumptions of constant hydraulic conductivity and constant hydraulic diffusivity, only the first term in the series need be considered. It can be shown that the effect of making allowance for the water table for the special case of constant K and constant D is to require the inclusion of the second term in the complete series solution.

For an initial constant rate of infiltration f_1 we can write equation (7.7) from Section 7.1 (recalling equations (7.21) and (7.44) for the change of variables) as

$$f_1 = K_1 \frac{\partial S}{\partial x} + K_1 \qquad (7.82a)$$

or in integrated form

$$S_1(x) = \left[1 - \frac{f_1}{K_1}\right](x_0 - x) \qquad (7.82b)$$

Since the moisture content is no longer constant in the profile, we must modify equation (7.62) given above and write it as

$$V(x, t) = K_2 \left[\frac{H + S_1(x) + x}{x}\right] \qquad (7.83)$$

Substitution from equation (7.82) into equation (7.83) gives

$$V(x,t) = K_2 \left[\frac{\{1 - \frac{f_1}{K_1}\}x_0 + H}{x} \right] + \frac{f_1}{K_1} K_2 \tag{7.84}$$

It will be noted that the second term on the right hand side of equation (7.84) depends on the rate of initial infiltration and will be zero if the soil column was in equilibrium before the start of infiltration. It could also be negative if the initial condition was one of capillary rise.

Because of the initial variation of moisture content, equation (7.63) must also be modified to give

$$f(t) = \left[c_2 - c_1(x) \right] \frac{dx}{dt} + f_1 \tag{7.85}$$

For the case of constant hydraulic conductivity K and constant hydraulic diffusivity D, the relationship between soil suction and soil moisture content will be given by equation (7.13) repeated here

$$S_1(x) = \frac{D}{K} [c_2 - c_1(x)] \tag{7.13}$$

By using equation (7.13) above and equation (7.82), equation (7.85) can be written as follows, for the case of constant K and constant D,

$$f(t) = \frac{K}{D} \left[1 - \frac{f_1}{K} \right] (x_0 - x) \frac{dx}{dt} + f_1 \tag{7.86}$$

If we take the depth of ponding H as small compared with the other terms in the numerator in equation (7.84), we have, for the special case of constant K and constant D, a particularly simple relationship, which is obtained by equating the percolation rate in the wetted zone given by equation (7.84) to the infiltration at the surface given by equation (7.86).

$$K \left[\frac{(1 - \frac{f_1}{K})x_0}{x} \right] + f_1 = \frac{K}{D} \left[1 - \frac{f_1}{K} \right] (x_0 - x) \frac{dx}{dt} + f_1 \tag{7.87a}$$

which simplifies to

$$Dx_0 = x(x_0 - x) \frac{dx}{dt} \tag{7.87b}$$

which integrates to give

$$\frac{D}{x_0^2} t = \frac{1}{2} \left[\frac{x}{x_0} \right]^2 - \frac{1}{3} \left[\frac{x}{x_0} \right]^3 \tag{7.88}$$

Since the above equation is dimensionless, it can be plotted as a single universal curve and used to find the relationship between the depth of penetration x and the elapsed time t in terms of the depth to the groundwater table x_0 and the hydraulic diffusivity of the soil D.

A second curve can be drawn on the same diagram giving the second dimensionless relationship.

$$\frac{f/K - f_1/K}{1 - f_1/K} = \frac{x_0}{x}$$

(7.89)

which is the special form of equation (7.84) for the assumptions made and enables us to relate the rate of infiltration f to the rate of initial infiltration f_1, the hydraulic conductivity of the soil K, and the depth of penetration x and hence to the elapsed time t. This relationship between the infiltration and the time elapsed will be given by

$$\frac{1}{2(\bar{f})^2} - \frac{1}{3(\bar{f})^3} = \frac{Dt}{x_0^2}$$

(7.90)

Dimensionless infiltration rate

where f is the dimensionless infiltration rate defined by

$$\bar{f} = \frac{f/K - f_1/K}{1 - f_1/K}$$

(7.91)

For the special case of $f_1 = K_1$, x_0 approaches infinity and equation (7.88) reduces to equation (7.66).

The above formulation has the advantage that it relates infiltration to the parameters that are of significance in soil moisture accounting in conceptual models of total catchment response. Thus, if the rain storm which produces flood runoff is preceded by some light precipitation at a rate less than the infiltration capacity of the soil, the assumption could be made that the initial rate of infiltration in the above equations f_1 was equal to the rate of antecedent precipitation. Alternatively, if the preceding period was one of net evapotranspiration, then the value of f_1 could be taken as minus the rate of the estimated evapotranspiration.

If we wish to model the total catchment response, we must be able to compute the recharge to the groundwater reservoir at the water table. For the classical Green–Ampt solution where a discontinuity at the water table is assumed, the recharge of the water table will be equal to the initial downward percolation rate f_1 until the wetting front reaches the water table. When this happens there will no longer be a suction ahead of

Wetting front

the wetting front. The depth of the wetted zone will be constant, so that equation (7.62) will become

$$r(t) = K_2 \left[\frac{x_0 + H}{x_0} \right]$$

(7.92)

The time during which the recharge to the water table will remain at the initial rate of f_1, before rising to the value of equation (7.92) can be obtained by substituting the value of the depth to the water table x_0 for the depth of penetration x in equation (7.65) above. For the model which allows for any rate of initial downward percolation to the water table (or upward capillary rise from it) but assumes constant values of K and D, the time during which the recharge at the water table (or loss of water from the water table) is given by equation (7.88) and the recharge after this time is given by equation (7.92) above.

If the high rate precipitation stops before the wetting front has reached the water table, then the analysis must be modified and the remaining time taken for the wetting front to reach the water table calculated on a new basis. For the Green–Ampt model of infiltration into a dry soil, it can be assumed that, following the end of precipitation and the infiltration of the ponded water, the surface layer will dry to the original condition so that the wetted zone will have the same suction at the top and the bottom. Under these circumstances a wetted zone of constant depth will travel downwards through the soil profile as a pulse at a constant rate equal to the saturated hydraulic conductivity. When the wetting front reaches the water table the recharge will instantaneously rise to a value equal to the saturated hydraulic conductivity but will afterwards decline because there will no longer be suction below the wetting front.

7.6 GROUNDWATER STORAGE AND OUTFLOW

Parallel field drains

There is a wide variety of groundwater conditions ranging from compact aquifers to karst topography. We will confine our attention here to the one-dimensional analysis of a simple case of groundwater flow. If we take the case where the land is drained by a set of parallel trenches, or parallel field drains, which are at a distance S apart, and which are subject to a constant rate of recharge r at the water table, the form of equation (7.6) given above for the equilibrium case will be

$$K \frac{\partial}{\partial h} h \left(\frac{\partial h}{\partial x} \right) + r = 0 \tag{7.93}$$

with the boundary conditions given by $h = d$ at both $x = 0$ and $x = S$, where d is the depth of water over the parallel drains, or the depth of water in the parallel trenches, whichever is appropriate. This is a non-linear equation, but because of its simple form an explicit solution can be found for the case examined. Integrating equation (7.93) once, we obtain

$$Kh \frac{\partial h}{\partial x} + rx = \text{constant} \tag{7.94a}$$

Since the first term of the left hand side of equation (7.94a) represents the horizontal discharge per unit width (see equation (7.4) in Section 7.1) and since by symmetry this discharge will be zero for a value of $x = S/2$, we can evaluate the constant in equation (7.94a)

$$Kh \frac{\partial h}{\partial x} + rx = \frac{rS}{2} \tag{7.94b}$$

The latter equation can once again be integrated with respect to x to give

$$\frac{Kh^2}{2} + \frac{r}{2}\left(x - \frac{S}{2}\right)^2 = \text{constant} \tag{7.95a}$$

Since K (the hydraulic conductivity), r (the rate of recharge) and S (the drainage spacing) are all constant, the above equation indicates that the shape of the water table profile between the drainage elements will take the form of an ellipse. If we take the water table depth as d in the neighbourhood of the drains and h_0 at the mid point between them, we can write

$$\frac{Kd^2}{2} + \frac{rS^2}{8} = \frac{Kh_0^2}{2} \tag{7.95b}$$

which enables us to determine any one of the parameters of interest when the others are known.

It must be remembered that equation (7.95) is based on the Dupuit–Forchheimer assumption. It is only correct if the flow can validly be approximated as a horizontal flow. If the drains or trenches do not penetrate to an impervious layer, or if the depth at the drains or trenches is small, this assumption may cease to be reasonable. However, it can be shown that even if the profile given on the basis of the Dupuit–Forchheimer assumptions is incorrect, the value of the discharge is correct. After all, this is what we are interested in, in hydrologic computations. Charny (1951) demonstrated mathematically in the case of two-dimensional seepage through a body of earth, with vertical upstream and downstream faces, and steady flow from a higher upstream body of water to a downstream body of water, that the lower level would be predicted exactly by the one-dimensional Dupuit–Forchheimer solution even though the profiles predicted in the two cases would be different. Aravin and Numerov (1953) extended this analysis to cover the case of seepage due to steady infiltration. For unsaturated flow, the Charny theorem does not hold but the errors are not large.

Two-dimensional seepage

The various solutions proposed for dealing with the problem as one of two-dimensional flow may be reviewed in such publications as Luthin (1957) and Kirkham (1966). For the case of a steady capillary rise from the groundwater to the surface, a similar analysis can be made to determine the shape of the drawdown in the water table between two parallel trenches set a distance (S) apart and each with a depth of water equal to d.

The basic equation (7.6) for the unsteady flow of groundwater in a horizontal direction was given in Section 7.1 above as

$$K\frac{\partial}{\partial x}\left(h\frac{\partial h}{\partial x}\right) + r(x, t) = f\frac{\partial h}{\partial t} \tag{7.6}$$

The above equation is non-linear and its solution for the unsteady case is quite difficult. Accordingly it is reasonable to consider what results can be obtained by linearisation of this basic equation. There are two ways in which equation (7.6) is usually linearised. In the first (and more common) linearisation, the height of the water table h in the first term of equation (7.6) can be frozen at some parametric value $\bar{h}$ and then removed outside the second differentiation with respect to x giving the linearised equation

$$K\bar{h}\frac{\partial^2 h}{\partial x^2} + r(x,t) = f\frac{\partial h}{\partial t} \tag{7.96}$$

which can be solved as a parabolic linear partial differential equation with constant coefficients. Since the equation is linear it can be solved for a delta function input or a step function input and the solution for a general input found from this basic input by convolution.

In the second form of linearisation, h^2 is used as the dependent variable instead of h and an equivalent parametric value of h is used to adjust the term on the right hand side of equation (7.6) in order to give

$$\frac{K}{2}\frac{\partial^2}{\partial x^2}(h^2) + r(x,t) = \frac{f}{2\bar{h}}\frac{\partial}{\partial t}(h^2) \tag{7.97}$$

Though the first linearisation given by equation (7.96) is more common, the second one given by equation (7.97) has the advantage that for the steady state it gives the correct non-linear solution represented by the ellipse equation derived in equation (7.95). It will be seen later that, while the two forms of linearisation give different shapes of profile, they give the same result for the groundwater outflow, which is the variable of interest in hydrology.

Transient behaviour The case of the transient behaviour of a groundwater reservoir drained by parallel ditches or field drains can be used to derive a model of groundwater flow. Equation (7.96) can be solved for this particular case subject to a delta function input by solving the equation

$$K\bar{h}\frac{\partial^2 h}{\partial x^2} = f\frac{\partial h}{\partial t} \tag{7.98}$$

for the initial condition of a level water table with the elevation of the water table above the impervious layer equal to h_0. The solution, which was obtained by Glover and included in Glover and Bittinger (1959), using the separation of variable technique, is given by

$$\frac{h-d}{h_0-d} = \frac{4}{\pi}\sum_{n=1,3,\ldots}^{\infty} n^{-1}\sin\left(\frac{n\pi x}{S}\right)\exp\left[-\frac{n^2\pi^2 K\bar{h}}{fS^2}t\right] \tag{7.99}$$

In the above solution, h is the elevation of the water table above the impervious layer, d is the elevation of the water surface in the trench (or above the field drain) above the impervious layer, h_0 is the maximum (i.e. the initial) elevation of the water table, x is the horizontal distance from the trench or drain, S is the spacing of the trenches or drains, K is the saturated permeability of the soil, f is the drainable pore space, and t is the time elapsed since the start of the recession.

Kraijenhoff van de Leur (1958) pointed out that the soil and drainage characteristics in equation (7.99) can be grouped into a single parameter, which he defined as the reservoir coefficient j which is given by

Reservoir coefficient

$$j = \frac{1}{\pi^2} \frac{fS^2}{K\bar{h}} \tag{7.100}$$

so that Glover's solution can be written as

$$\frac{h-d}{h_0-d} = \frac{4}{\pi} \sum_{n=1,3,\dots}^{\infty} n^{-1} \sin\left(\frac{n\pi x}{S}\right) \exp\left(-n^2 \frac{t}{j}\right) \tag{7.101}$$

Kraijenhoff (1958) also pointed out that Glover's solution was the solution for finite volume of recharge in an infinitesimal time and consequently that equation (7.101) represents the impulse response of the groundwater system.

If we adopt the second linearisation instead of the first, a similar equation can be obtained except that it will be in terms of h^2 rather than h. The difficulty about conflicting predictions of the shape of the water table profile, does not affect us in our study of the recession of outflow.

The outflow to a trench or a drain is given by evaluating the discharge at $x = 0$ and $x = S$ and combining the two values to give the total outflow. When using the first linearisation the discharge at any point is given by

$$q = -K\bar{h}\frac{\partial h}{\partial x} \tag{7.102}$$

By substituting the solution given by equation (7.101) in equation (7.102) and combining the results of $x=0$ and $x=S$ we obtain for the total outflow

$$q = \frac{8K\bar{h}(h_0-d)}{S} \sum_{n=1,3,\dots}^{\infty} \exp\left(-n^2 \frac{t}{j}\right) \tag{7.103}$$

If the problem had been solved by means of the second linearisation, the discharge would have been given by

$$q = -K\frac{\partial}{\partial x}(h^2) \tag{7.104}$$

and on substitution of the solution in terms of h^2 in this equation and evaluating the discharges, the same results would be obtained as in equation (7.103).

Instantaneous recharge

For an initial height of instantaneous recharge of $(h_0 - d)$ the volume of recharge can be expressed in terms of this height, the drain spacing and the drainable porosity of the soil as

$$\forall = (h_0 - d)Sf \tag{7.105}$$

so that in terms of the volume of recharge the outflow is given by

$$q = \frac{8K\bar{h}}{S^2 f} \forall \sum_{n=1,3,\ldots}^{\infty} \exp\left(-n^2 \frac{t}{j}\right) \tag{7.106}$$

Consequently, for an instantaneous input of unit volume we have as the impulse response

$$h(t) = \frac{8}{\pi^2} \frac{1}{j} \sum_{n=1,3,\ldots}^{\infty} \exp\left(-n^2 \frac{t}{j}\right) \tag{7.107}$$

Recession curve

Obviously, when it becomes large, the first term in the infinite series in the above equation will dominate and the shape of the recession curve will approximate that from a single linear reservoir. For small values of t, however, the contribution of the other terms cannot be neglected, and as t approaches zero they each approach unity and their sum approaches an infinite value.

Equation (7.107) above represents a one-parameter conceptual model and this is fitted easily to field data. The response function has as the first cumulant or first moment about the origin

$$k_1 = U_1' = \frac{\pi^2}{12} j \tag{7.108}$$

and for its second and third cumulants (i.e. second and third moments about the centre) the values

$$k_2 = U_2 = \frac{7\pi^4}{720} j^2 \tag{7.109}$$

$$k_3 = U_3 = \frac{31\pi^6}{15,120} j^3 \tag{7.110}$$

Glover–Kraijenhoff model

Accordingly, the shape factors for the Glover–Kraijenhoff model represented by equation (7.107) are given by

$$s_2 = \frac{7}{5} \tag{7.111}$$

$$s_3 = \frac{124}{35} \tag{7.112}$$

Even though this particular conceptual model approximates a single linear reservoir at high values of elapsed time t, it can be simulated more closely (if moment matching is taken as the criterion) by a cascade of linear reservoirs with a value of $n = 0.7$ and a value of $K = 1.15$.

Applied hydrologists frequently assume that the recession for the groundwater phase of catchment response is given by an exponential decline:

$$Q = Q_0 \exp(-at) \tag{7.113}$$

and attempt to verify this assumption by plotting the recession part of the hydrograph on semi-log paper. It is clear that this approach not only assumes that the groundwater reservoir acts in linear fashion, but assumes that it behaves in the same way as a particularly simple linear system, namely a single linear reservoir. If we are prepared to make this assumption for the recession phase of the groundwater outflow, we should try to extract as much as we can from the assumption by assuming that the recharge can be simulated by the same simple model. If such a system is subject to a recharge at a uniform rate R, then the groundwater outflow during recharge will be given by

$$Q(t) = R[1 - \exp(-at)] + Q_0 \exp(-at) \tag{7.114}$$

where the time origin is taken at the start of recharge. If the recharge ends after a time D, the groundwater outflow at this time will be given by

$$Q(D) = R[1 - \exp(-aD)] + Q_0 \exp(-aD) \tag{7.115}$$

and the recession from this time on will be given by

$$Q(t) = Q(D) \exp(-a(t - D)) \tag{7.116}$$

It is clear that the recession after recharge, represented by equation (7.116), is the same as would have occurred, if there had been an instantaneous increase in discharge at a time $t = 0$ of an amount

$$\Delta Q = R[\exp(aD) - 1] \tag{7.117}$$

The above property indicates that a semi-log plot of the hydrograph and the knowledge of the volume of recharge might enable us to determine the rate and duration of recharge. Apart from this possible method of analysis, equation (7.114) indicates that the separation between groundwater and direct storm runoff for this simple model should be taken as a curve which is concave downwards rather than as a straight line as is usually done.

Recession of groundwater outflow
　　　　Other simple models can be devised for the recession of groundwater outflow. Thus we could estimate outflow on the basis of a succession of steady states in each of which the non-linear solution represented by the ellipse equation was assumed. The relationship between discharge and the depth of water at the drain and mid-way between them is on the basis of equation (7.95) above given by

$$Q = rS = \frac{4K}{S} \left(h_{max}^2 - d^2 \right) \tag{7.118}$$

For the case where rise of water table is very small compared to the depth of water in the drainage trench, or over the field drain, we can write the discharge as

$$Q = \frac{8Kd}{S}(h_{max} - d) \tag{7.119}$$

and the volume of storage above the level in the drainage trench as

$$\forall = f\frac{rS^3}{12Kd} = \frac{2}{3}fS(h_{max} - d) \tag{7.120}$$

Comparison of equation (7.119) and equation (7.120) indicates that the volume is proportional to the outflow, so that we have the case of a single linear reservoir, whose storage delay time is given by equation

$$K = \frac{\forall}{Q} = \frac{1}{12}\frac{fS^2}{Kd} = \frac{\pi^2}{12}j \tag{7.121}$$

so that for a very shallow rise of water table the transient behaviour can be represented by a single linear reservoir, whose storage delay time is about three per cent smaller than the reservoir coefficient of the groundwater reservoir itself. In the other limiting case, where the depth in the drainage trench is small in comparison with the rise of the water table, we can write the discharge as

$$Q = \frac{4K}{S}h^2_{max} \tag{7.122a}$$

$$\forall = f\frac{\pi Sh_{max}}{4} \tag{7.122b}$$

so that the discharge will be proportional to the square of volume.

$$Q = \frac{64K}{f^2\pi^2S^2}(\forall)^2 \tag{7.123}$$

Non-linear reservoir

For intermediate conditions, it might be possible to simulate the recession of groundwater outflow with fair accuracy by treating the groundwater system as a non-linear reservoir with the outflow proportional to some power of storage between 1 and 2.

CHAPTER 8

Non-linear Deterministic Models

8.1 NON-LINEARITY IN HYDROLOGY

Non-linear methods

If we examine the basic physical equations governing the various hydrologic processes, we find that these equations (and hence the processes they
represent) are non-linear. Consequently, we face the distinct possibility
that all of the approaches of linear analysis discussed in Chapters 4, 5, 6
and 7 may be irrelevant to real hydrologic problems, save as a prelude to
the development of non-linear methods. Accordingly, in the present chapter we take up this question of non-linearity and ask ourselves whether
we can determine under what circumstances the effects of non-linearity
will be most marked and also whether we can adapt the methods of linear analysis described in previous chapters to the non-linear case. While
knowledge of linear methods of analysis is valuable in such an examination, we must avoid the tendency to carry over into non-linear analysis
certain preconceptions, which are valid only for the linear case.

The basic equations for the one-dimensional analysis of unsteady
flow in open channels are the continuity equation and the equation for
the conservation of linear momentum. The continuity equation can be
written as:

$$\frac{\partial Q}{\partial x} + \frac{\partial A}{\partial t} = r(x, t) \tag{8.1}$$

where Q is the discharge, A the area of flow, and $r(x, t)$ the rate of lateral
inflow. The above equation is a linear one and consequently poses no
difficulties for us in this regard. The second equation used in the one-
dimensional analysis of unsteady free-surface flow is that based on the
conservation of linear momentum, which reads

$$\frac{\partial y}{\partial x} + \frac{u}{g}\frac{\partial u}{\partial x} + \frac{1}{g}\frac{\partial u}{\partial t} = S_0 - S_f - \frac{u}{gy}r(x, t) \tag{8.2}$$

where y is the depth of flow, u is the mean velocity, S_0 is the bottom
slope and S_f is the friction slope. This dynamic equation is highly non-
linear. Consequently, it is not possible to obtain closed-form solutions
for problems governed by equations (8.1) and (8.2). The extent of the

non-linearity can be appreciated if we examine the special case of discharge in an infinitely wide channel with Chezy friction, in which case the continuity equation takes the form

$$\frac{\partial q}{\partial x} + \frac{\partial y}{\partial t} = r(x, t) \tag{8.3}$$

where $q = uy$ is the discharge per unit width; and the momentum equation takes the form

$$\frac{\partial y}{\partial x} + \frac{u}{g}\frac{\partial u}{\partial x} + \frac{1}{g}\frac{\partial u}{\partial t} = S_0 - \frac{u^2}{C^2 y} - \frac{u}{gy}r(x, t) \tag{8.4a}$$

which appears to be non-linear in only three of its six terms. However, if we multiply through by gy, five of the six terms of the equation are seen to be non-linear. If, in addition, we express u in terms of q and y, which are the dependent variables in the linear continuity equation, we obtain

$$(gy^3 - q^2)\frac{\partial y}{\partial x} + 2qy\frac{\partial q}{\partial x} + y^2\frac{\partial q}{\partial t} = S_0 gy^3 - \frac{g}{C^2}q^2 \tag{8.4b}$$

in which every term is seen to be highly non-linear (see Appendix D).

On the basis of the above equations we would expect such processes *Unsteady flow with a* as flood routing, which is a case of unsteady flow with a free surface, to *free surface* be characterised by highly non-linear behaviour. However, practically all the classical methods of flood routing commonly used in applied hydrology are linear methods. In contrast most of the methods used in applied hydrology to analyse overland flow (which is another case of unsteady free surface flow) are non-linear in character.

Sub-surface flow The basic equations for sub-surface flow are also non-linear in form. For the case of one-dimensional unsteady vertical flow in the unsaturated zone, the basic equation (often known as Richards equation) was developed in Section 7.1 above and given as equation (7.11) on page 129. This equation reads, in its diffusivity form, as

$$\frac{\partial}{\partial z}\left[D(c)\frac{\partial c}{\partial z}\right] + \frac{\partial}{\partial z}[K(c)] = \frac{\partial c}{\partial t} \tag{8.5}$$

where c is the moisture content, $K(c)$ the hydraulic conductivity of the unsaturated soil at a moisture content c, and $D(c)$ the hydraulic diffusivity of the unsaturated soil at a moisture content c. Since the hydraulic conductivity is usually a non-linear function of c and the hydraulic diffusivity varies with c, equation (8.5) above is clearly a non-linear equation and so represents a non-linear process.

The equation for horizontal flow in the saturated zone was also derived in Section 7.1 as equation (7.6) on page 128 and is usually known as the *Boussinesq equation* Boussinesq equation:

$$K\frac{\partial}{\partial x}\left(h\frac{\partial h}{\partial x}\right) + r(x, t) = f\frac{\partial h}{\partial t} \tag{8.6}$$

where h is the height of the water table above a horizontal impervious layer, K is the saturated hydraulic conductivity, f the drainable porosity of the soil, and $r(x, t)$ is the rate of recharge at the water table. This equation is clearly non-linear, because of the first term on the left-hand side of the equation, and also because the usual assumption that f may be taken as constant, is open to serious doubt.

There are so many uncertainties in the derivation of unit hydrographs that reliable and significant data on the existence of non-linearity in surface response is not readily available. However, there have been some interesting results which have been published in the literature and which need to be taken into account in any attempt to evaluate the non-linearity of catchment response. Minshall (1960) derived unit hydrographs for a small catchment of 27 acres in Illinois for five storms whose average intensity varied from 0.95 inches per hour to 4.75 inches per hour. If this small catchment acted in a linear fashion, the unit hydrographs should have been essentially the same for each of the five storms.

Non-linearity of catchment response

Actually, as shown in Figure 8.1 the peak of the unit hydrograph showed a more than threefold variation, being higher for the greater rainfall intensity. The time to peak also showed a threefold variation, being smaller for the larger rainfall intensities. Minshall's results are a clear indication of non-linear behaviour.

Amorocho and Orlob (1961) and Amorocho and Brandsetter (1971) published data for a very small laboratory test basin, in which the artificial rainfall was carefully controlled and the runoff accurately measured (see

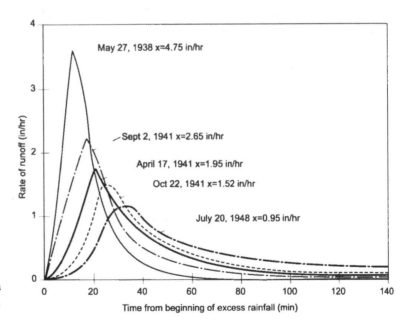

Figure 8.1. Hydrograph for a 27 acre catchment (Minshall, 1960).

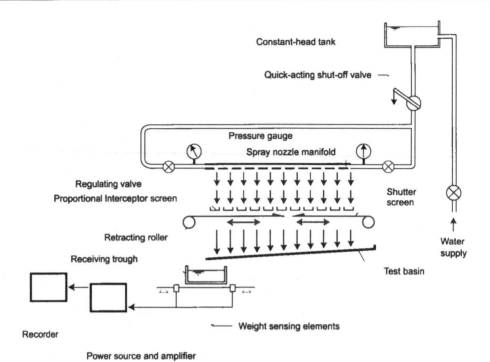

Figure 8.2. Laboratory experiment (Amorocho and Brandstetter, 1971).

Figure 8.2). The test basin consisted of a thin layer of gravel placed over an impervious surface. The results for varying rates of input showed a clearly non-linear response as indicated by the set of three experimental results, in which the cumulative outflows are not proportional to the corresponding cumulative inflows (see Figure 8.3). Both Minshall's and Amorocho's data will be discussed later in Section 8.5, which deals with the concept of spatially uniform non-linearity.

Ishihara and Takasao (1963), in a paper on the applicability of unit hydrograph methods, showed results for the Yura river basin at Ono, which is a river basin of 346 square kilometres. Figure 8.4 presents the relationship, which they obtained, between the mean rainfall intensity and the time of rise between the start of the equivalent mean rainfall and the peak of the flow hydrograph. They interpreted the events for a mean rainfall intensity above 10 millimetres per hour as representing essentially surface runoff, and the events for a mean rainfall intensity of less than 8 millimetres per hour as representing essentially sub-surface runoff. It is noteworthy, that their results, as presented in Figure 8.4, show remarkably little variation in the time to peak for rainfall intensities from 4 millimetres per hour to 18 millimetres per hour, but show a distinct variation for smaller rainfall intensities. The results from Ishihara and Takasao also indicate a linear

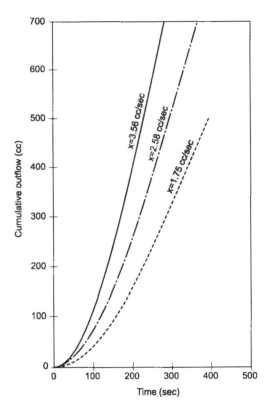

Figure 8.3. Typical results of laboratory experiment (Amorocho and Orlob, 1961).

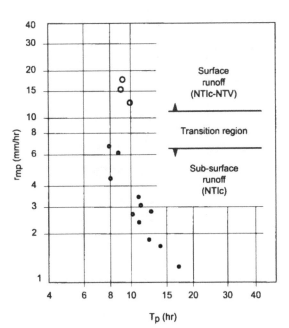

Figure 8.4. Time to peak versus rainfall intensity (Yura river) (Ishihara and Takasao, 1963).

relationship between peak runoff and mean intensity of rainfall at higher values of rainfall intensity. These two results taken together would indicate that for conditions similar to those in the Yura river catchment, the unit hydrograph approach might be reliable for high intensities, but not for low ones.

Time of particle travel   Pilgrim (1966) measured the time of particle travel in a catchment area of 96 square miles by means of radioactive tracers. His results showing the relationship between time of travel and level of discharge are presented in Figure 8.5. As in the case of the Japanese result, we see here an essential constancy of time of travel at higher rates of discharge and a tendency for the time of travel to be inversely proportional to the discharge at lower discharges.

We conclude from this brief summary, both from the basic equations of physical hydrology and from experimental data, that there are sufficient indications of non-linearity to justify an investigation of the extent to which non-linearity affects the techniques commonly used in applied hydrology.

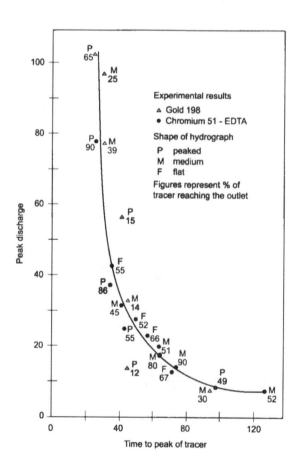

Figure 8.5. Time of travel versus discharge (Pilgrim, 1966).

There are many possible approaches to the analysis of non-linear processes and systems. One approach is to analyse each input–output event as if it were linear and then to examine the effects of the level of input on the results obtained. Linearisation can be applied to all three basic approaches used in hydrology: black-box analysis, conceptual models, or solution of the basic equations. The Linearisation approach is discussed in Section 8.3 below.

Non-linear method of black-box analysis

A second line of approach to non-linear systems is to accept the non-linearity and to attempt to develop a non-linear method of black-box analysis, which would be a generalisation of linear black-box analysis as discussed in Chapter 4. This method is summarised in Section 8.4. Still another approach would be to attempt to find simple non-linear conceptual models, which would simulate the operation of non-linear systems with the same degree of accuracy as achieved by the simple linear conceptual models presented in Chapter 5. This is the subject of Sections 8.2 and 8.5. A final approach would be to accept the full complexity of the complete non-linear equation and to seek solutions by numerical methods. This last approach is outside the scope of this book.

8.2 THE PROBLEM OF OVERLAND FLOW

Inherent non-linearity

Overland flow is an interesting example of a hydrologic process, which appears to require a non-linear method of solution. It would appear that because it occurs early in the runoff cycle, the inherent non-linearity of the process is not dampened out in any way as appears to occur to some extent in the question of catchment runoff. A physical picture of overland flow is shown in Figures 8.6 and 8.7 together with a few of the classical experimental results of Izzard (1946).

For the two-dimensional problem of lateral inflow the equation of continuity is written as

$$\frac{\partial q}{\partial x} + \frac{\partial y}{\partial t} = r(x, t) \tag{8.7}$$

where q is the rate of overland flow per unit width, y is the depth of overland flow and r is the rate of lateral inflow per unit area. The equation for the conservation of linear momentum is written as (from equation 8.4a)

$$\frac{\partial y}{\partial x} + \frac{u}{g}\frac{\partial u}{\partial x} + \frac{1}{g}\frac{\partial u}{\partial t} = S_0 - S_f - \frac{q}{gy^2}r(x, t) \tag{8.8}$$

where u is the velocity of overland flow, S_0 is the slope of the plane and S_f is the friction slope.

The classical problem of overland flow is the particular case where the lateral inflow is uniform along the plane and takes the form of a unit step function. There are several parts to the complete solution of this problem.

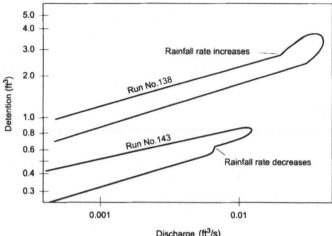

Figure 8.6. Storage versus
discharge (Izzard, 1946).

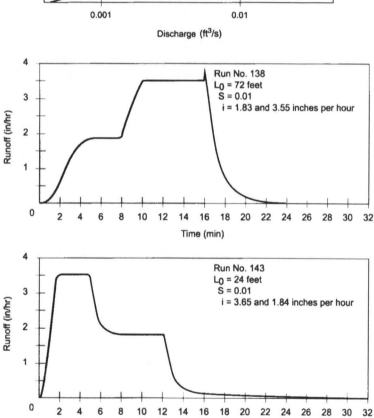

Figure 8.7. Hydrograph of
overland flow (Izzard, 1946).

Firstly, there is the steady-state problem of determining the water surface
profile when the outflow at the downstream end of the plane increases
sufficiently to balance the inflow over the surface of the plane. Secondly,
there is the problem of determining the rising hydrograph of outflow before

this equilibrium state is approached for the special case of the step function input. If the process were a linear one, the solution of this second problem (i.e. the determination of the step function response) would be sufficient to characterise the response of the system and the outflow hydrograph, for any other inflow pattern, could be calculated from it.

However, since the problem is inherently non-linear, the principle of superposition cannot be used and each case of inflow must be treated on its merits. The third part of the classical problem is that of determining the recession from the equilibrium condition after the cessation of long continued inflow. Further problems that must be investigated are the nature of the recession when the inflow ceases before equilibrium is reached, the case where there is a sudden increase from one uniform rate of inflow to a second higher uniform rate of inflow, and the case when a uniform rate of inflow is suddenly changed to a second rate of uniform inflow, which is smaller than the first. The above problems can be solved by numerical methods (Liggett and Woolhiser, 1967; Woolhiser, 1977) but such methods are outside the scope of the present discussion. Here we will be concerned with simpler approaches to the problem and with attempts to find a simple mathematical simulation or a simple conceptual model.

The first approach to the solution of overland flow in classical hydrology was based on the replacement of the dynamic equation, given by equation (8.8) above, by an assumed relationship between the outflow at the downstream end of the plane and the volume of storage on the surface of the plane. Because this method was first proposed by Horton (1938) for overland flow on natural catchments and subsequently used by Izzard (1946) for impermeable plane surfaces, it may be referred to as the Horton–Izzard approach. It had been noted by hydrologists that for equilibrium conditions on experimental plots, the relationship between the equilibrium runoff and the equilibrium storage could be approximated by a power relationship. Such a relationship would indicate that the outflow and the storage would be connected as follows

$$q(L, t_e) = q_e = a(S_e)^c \tag{8.9}$$

where q_e is the equilibrium discharge at the downstream end ($x = L$) after a lapse of time t_e sufficient for equilibrium to occur, S_e is the total surface storage at equilibrium conditions, and a and c are parameters which could be determined from experimental data by means of a log–log plot.

Recession

In the Horton–Izzard approach to the overland flow problem, the assumption is made that such a power relationship holds, not only at equilibrium, but also at any time during the period of unsteady flow, either during the rising hydrograph or during the recession. This assumption can be written as

$$q(L, t) = q_L = a(S)^c \tag{8.10}$$

where q_L is the discharge at the downstream end at any time t and S is the corresponding storage on the surface of the plane of overland flow at the same time. Izzard (1946) illustrates the nature of this approximate relationship on Figure 8.7 for two of the experimental cases examined.

The equation of continuity in its lumped form for the whole plane can be written for the case of constant input r as

$$\frac{dS}{dt} = rL - q_L \tag{8.11}$$

which is in reality an integrated form of equation (8.7). If the Horton–Izzard assumption given in equation (8.10) is made, then this equation of continuity can be written as

$$\frac{dS}{dt} = q_e - aS^c \tag{8.12a}$$

or in more convenient form as

$$dt = \frac{dS}{q_e - aS^c} \tag{8.12b}$$

Equation (8.12) can be integrated to give the time as a function of the storage

$$t = \frac{S_e}{q_e} \int \frac{d(S/S_e)}{1 - (S/S_e)^c} \tag{8.13}$$

Equation (8.13) can be solved analytically for values of $c = 1$ (linear case), $c = 2$, $c = 3$ or $c = 4$ and also for values of c which are ratios of these integral values i.e. for $c = 3/2$ or $c = 4/3$. It is interesting to note that the integral in equation (8.13) occurs also in the case of non-uniform flow in an open channel (Bakmeteff, 1932) and in the case of the relationship between actual and potential evapo-transpiration (Bagrov, 1953; Dooge, 1991).

Rising hydrograph   Horton (1938) solved the equation of the rising hydrograph given by equation (8.13) for the case of $c = 2$, which he described as "mixed flow" since the value of c is intermediate between the value of 5/3 for turbulent flow and the value of 3 for laminar flow. If we define a time parameter K_e by

$$K_e = \frac{S_e}{q_e} = \frac{1}{aS_e^{c-1}} = \frac{1}{a^{1/c}q_e^{(c-1)/c}} \tag{8.14}$$

the integration of equation (8.13) with $c = 2$ for zero initial condition gives

$$\frac{2t}{K_e} = \log_e \left[\frac{1 + (S/S_e)}{1 - (S/S_e)} \right] \tag{8.15a}$$

which gives the time as a function of the storage. This can readily be rearranged to give the storage as an explicit function of time:

$$\frac{S}{S_e} = \tanh\left[\frac{t}{K_e}\right] \tag{8.15b}$$

and hence to give the discharge as an explicit function of time

$$\frac{q}{q_e} = \tanh^2\left[\frac{t}{K_e}\right] \tag{8.15c}$$

Since the system is non-linear, the time parameter K_e will depend on the intensity of the inflow. Horton's equation as given above has been widely used in the design of airport drainage systems in the period since he proposed it almost forty year ago. The solution of equation (8.13) for the case of $c = 3$ i.e. for laminar flow was presented by Izzard (1944) in the form of a dimensionless rising hydrograph. Izzard, who appears to have followed the theoretical analysis of Keulegan (1944), uses as his time parameter a time to virtual equilibrium, which is defined as twice the time parameter given in equation (8.14) above.

For recession from equilibrium, the recharge in equation (8.11) becomes zero, and the substitution for q_L from equation (8.10) and a slight rearrangement gives us the simple differential equation

$$-a\,dt = \frac{dS}{S^c} \tag{8.16}$$

which can be solved for any value of c. For the linear case ($c = 1$) the storage and hence the outflow shows an exponential decline. For all other values of c the recession of equilibrium is given by

$$\frac{q}{q_e} = \frac{1}{\left[1 + (c-1)(t/K_e)\right]^{c/(c-1)}} \tag{8.17}$$

where q is the outflow at a time t after the start of recession i.e. after the cessation of inflow. The special case of equation (8.17) for laminar flow (i.e. for $c = 3$) was given by Izzard (1944). Figure 8.6 shows the rising hydrograph and the recession for the Horton–Izzard solution with $c = 2$ for a duration of inflow equal to twice the time parameter defined by equation (8.14). The double curvature of the rising hydrograph is characteristic of the shape of the rising hydrograph for the Horton–Izzard solution for all values of c other than $c = 1$. If the duration of inflow D is less than the *Partial recession* time required to reach virtual equilibrium, we get a partial recession from the value of the outflow q_D which has been reached at the end of inflow. It can be shown that the partial recession curve has the same shape as the recession from equilibrium given by equation (8.17), except that the recession curve from partial equilibrium starts at the point on the curve defined by the appropriate value of q_D/q i.e. the end of inflow does not correspond to the time origin in equation (8.17).

If there is a change to a new rate of uniform inflow during the rising hydrograph one of two cases can occur. If the new rate of inflow is higher than the rate of outflow when the change occurs, the remainder of the rising hydrograph will follow the same dimensionless curve as before; but since q_e is equal to the inflow at equilibrium, the value of q/q_e will change as soon as the rate of inflow changes. Such a case is shown in Figure 8.7a (run No. 138). If the new rate of inflow is less than the outflow at the time when the change occurs, the hydrograph will correspond to the general falling hydrograph for the case in question. An example of such a falling hydrograph taken from Izzard (1944) is shown on Figure 8.7b (run No. 143). For the case of $c = 2$, the equation for the falling hydrograph can also be obtained. For the case of recession to equilibrium, the integration given in equation (8.15a) above is not valid, because it results in the logarithm of a negative number. However, it can be shown that the appropriate integration in this case is

$$\frac{2t}{K_e} = \log_e \left[\frac{S/S_e + 1}{S/S_e - 1} \right] \tag{8.18a}$$

which can be rearranged to give the discharge as an explicit function of time:

$$\frac{q}{q_c} = \coth^2 \left[\frac{t}{K_c} \right] \tag{8.18b}$$

In practice, it would often be convenient to take advantage of the relationship between tanh and coth and to write

$$\frac{q_e}{q} = \tanh^2 \left[\frac{t}{K_e} \right] \tag{8.18c}$$

so that one could use either a single graph, or a single computer routine, to evaluate the value of q/q_e for the rising hydrograph, or the value of q_e/q for the falling hydrograph.

Master recession curve   The master recession curve defined by equation (8.18) applies to all cases where there is a uniform rate of lateral inflow and an initial storage on the plane which is higher than the equilibrium storage for that particular inflow. There will be a similar master recession curve for any other value of the index of non-linearity c. The only case to which this master recession curve will not apply is when the inflow drops to zero. In the latter case, the governing equation will be equation (8.16) and if allowance is made for the discharge q_0 at the cessation of inflow, we will have the relationship

$$[q_e/q]^{(c-1)/c} - [q_e/q_0]^{(c-1)/c} = (c - 1)\frac{t}{K_e} \tag{8.19a}$$

which can be rearranged to give the discharge as a function of time

$$\frac{q}{q_e} = \frac{1}{\left[[q_e/q_0]^{(c-1)/c} + (c-1)(t/K_e)\right]^{c/(c-1)}} \tag{8.19b}$$

of which equation (8.17) is a special case.

The Horton–Izzard approach as described above clearly involves the use of a simple conceptual model. In fact, the whole approach is based on treating the overland flow as a lumped non-linear system, which can be represented by a single non-linear reservoir whose operation is described by equation (8.10) above. The Horton–Izzard solution for the various cases arises from the application of the equation of continuity to such a single non-linear reservoir. Even though this conceptual model is extremely simple in form, the fact that it is non-linear makes it less easy to handle than some of the apparently more complex conceptual models used to simulate linear or linearised systems. Thus the impulse response for such a system no longer characterises the operation of the system because the output in any individual case will depend on the form and intensity of the input. The cumulants of the impulse response can no longer be added to the cumulants of the input to obtain the cumulants of the output. The solution for a step function input (described above as the rising hydrograph solution) cannot be used to predict the output for a varying pattern of input. However, because the method involves only a single non-linear reservoir the output for a complex pattern of input can be obtained by dividing the input into intervals of uniform inflow and applying to each interval one or other of the three master solutions (the rising hydrograph, the recession to equilibrium, the recession to zero) for the chosen value of the index of non-linearity c.

Non-linear reservoir

The second simplified approach to the problem of overland flow, which appears in the literature, is the kinematic wave solution (Henderson and Wooding, 1964). This also involves a power relationship between discharge and depth. In this case the relationship is not a lumped one for the whole system, but a distributed relationship between the discharge and the depth at each point. This basic relationship can be written as

Kinematic wave solution

$$q(x,t) = b[y(x,t)]^c \tag{8.20}$$

where $q(x,t)$ is the discharge per unit width, $y(x,t)$ is the depth of the flow, and a and c are parameters of the system. The above basic assumption corresponds to neglecting all of the terms in the equation for the conservation of linear momentum given by equation (8.8) above in comparison with the terms for the bottom slope and the friction slope, so that we can write

$$S_0 - S_f = 0 \tag{8.21}$$

and also make the further assumption that the friction may be evaluated by a power law. If friction is taken according to the Chezy formula, the value of c in equation (8.20) will be 3/2, whereas if it is taken according to the Manning formula, the value of c will be 5/3.

As might be expected, the solution of the basic problems of overland flow for this distributed model cannot be derived as easily as the solution for the lumped model represented by the Horton–Izzard approach. The problem is to solve equations (8.7) and (8.20) subject to the appropriate boundary conditions. The method of characteristics can be used (Eagleson, 1969; Woolhiser, 1977) to obtain the solution, but the details of the derivation are outside the scope of the present chapter. For the case of the rising hydrograph it can be shown that the solution is given by

Method of
characteristics

$$\frac{q}{q_e} = \left[\frac{t}{t_k}\right]^c \tag{8.22}$$

from $t = 0$ to $t = t_k$ where the latter time parameter is given by

$$t_k = \frac{y_e}{r} = \frac{1}{r}\left[\frac{q_e}{b}\right]^{1/c} \tag{8.23}$$

where q_e is the equilibrium discharge at the downstream end of the overland slope and y_e is the depth of flow at the downstream end for equilibrium conditions and small b is the parameter in equation (8.20) above. For times greater than the kinematic time parameter, the discharge is equal to the equilibrium discharge. The recession from full equilibrium for the kinematic wave solution can be shown to be

$$\frac{1 - q/q_e}{(q/q_e)^{(c-1)/c}} = c\left[\frac{t}{t_k}\right] \tag{8.24}$$

where t is the time since the cessation of inflow.

Care should be taken to distinguish the *kinematic time parameter t_k* from the *time parameter for the Horton–Izzard model K_e* defined by equation (8.14) above. The relationship between the two time parameters can be demonstrated as follows. The storage at equilibrium is given by

Horton–Izzard model

$$S_e = \int_0^L y_e(x)\,dx \tag{8.25}$$

where y_e is the equilibrium depth at a distance x from the upstream end. Since equation (8.20) holds at equilibrium, this can be written

$$S_e = \int_0^L \left[\frac{q_e(x)}{b}\right]^{1/c}\,dx \tag{8.26}$$

At equilibrium we have

$$q_e(x) = rx \tag{8.27}$$

so that equation (8.26) can be written

$$S_e = \int_0^L \left(\frac{rx}{b}\right)^{1/c}\,dx \tag{8.28}$$

The latter equation can be integrated to give

$$S_e = \left(\frac{c}{c+1}\right)\left(\frac{r}{b}\right)^{1/c} (L)^{(c+1)/c} \tag{8.29a}$$

which can be regrouped as

$$S_e = \left(\frac{c}{c+1}\right)\left(\frac{rL}{b}\right)^{1/c} L \tag{8.29b}$$

which by virtue of equations (8.23) and (8.27) is

$$S_e = \left(\frac{c}{c+1}\right) t_k q_e \tag{8.29c}$$

Recalling the definition of the Horton–Izzard time parameter as given by equation (8.14) we see that

$$K_e = \frac{S_e}{q_e} = \left(\frac{c}{c+1}\right) t_k \tag{8.30}$$

which is the required relationship between the time parameter for the Horton–Izzard approach and the kinematic time parameter.

The only sound basis for evaluating the above two models is by comparing the results with solutions of the complete non-linear equation. The numerical solution of the overland flow problem has been tackled by Woolhiser and Liggett (1967). They reduce the equations of continuity and momentum to dimensionless forms by expressing variables in terms of the normal depth and the normal velocity at the downstream end of the plane for the equilibrium discharge. When this is done, the continuity equation given above by equation (8.7) becomes

$$\frac{\partial q}{\partial x} + \frac{\partial y}{\partial t} = 1 \tag{8.31}$$

where q is the ratio of the discharge to the equilibrium discharge at the downstream end and y is the ratio of the depth to the equilibrium depth at the downstream end of the plane. The equation for linear momentum becomes

$$\frac{\partial y}{\partial x} + F_0^2\left(u\frac{\partial u}{\partial x} + \frac{\partial u}{\partial t} + \frac{u}{y}\right) = \frac{S_0 L}{y_0}\left[1 - \frac{u^2}{y}\right] \tag{8.32}$$

where u is the ratio of the velocity to the equilibrium velocity at the downstream end and F_0 is the Froude number for the equilibrium flow at normal depth. In both equations (8.31) and (8.32) the independent variable x is the ratio of the distance from the upstream end to the total length of overland flow. The independent variable t is the ratio of the time to the characteristic time t_0 obtained by dividing the length of the plane by the velocity at the downstream end at equilibrium. This reference time t_0 can be shown to be equal to the kinematic time parameter defined by

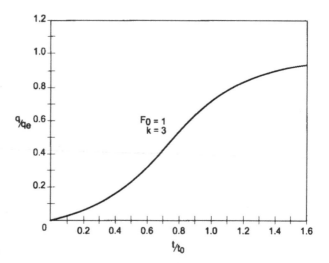

Figure 8.8. Overland flow
(Ligget and Woolhiser, 1967).

Dimensionless length

equation (8.23).It appears from these dimensionless equations that there are only two parameters governing the flow, namely the Froude number for normal flow at equilibrium discharge F_0 and the dimensionless length factor $S_0 L/y_0$. When the ratio of these two parameters defined by

$$K = \frac{S_0 L}{F_0^2 y_0} \tag{8.33}$$

is appreciably greater than 1, all of the terms on the left-hand side equation (8.32), except the first, can be neglected, thus reducing the momentum equation to convective-diffusion form. If in addition the dimensionless length factor $S_0 L/y_0$ is also appreciably greater than 1, the first term on the left hand side of equation (8.32) can be neglected as well and the kinematic wave approximation results.

Woolhiser and Liggett (1967) found that for values of K as defined by equation (8.33) greater than 10, the kinematic wave solution was a good approximation to the rising hydrograph, but that for values of K smaller than 10 the approximation was a poor one. Figure 8.8 shows a solution obtained by Ligett and Woolhiser for a value of the Froude number equal to 1.0 and a value of the parameter K equal to 3.0.

8.3 LINEARISATION OF NON-LINEAR SYSTEMS

If we know the input and the output for a non-linear system, there is nothing to prevent us using the techniques discussed in Chapter 4 to obtain the apparent unit hydrograph of the system. If we convolute the "unit hydrograph" thus derived with the input, we should obtain an accurate reconstitution of the output. The difficulty is that for the non-linear system, we cannot use such a derived "unit hydrograph" to predict by direct convolution the output for any other input. Only in the case of a linear time-invariant system is the unit hydrograph obtained by the solution of the convolution equation independent of the particular event from which it is derived. However, if we can derive such an apparent unit hydrograph from a number of storms with markedly different inputs of effective precipitation, we might be able to make progress. We seek information on the manner in which key parameters (the time to peak and the peak discharge, or the values of the lower moments) of the apparent unit hydrograph vary with the characteristics of the input. And, we may be able to predict the shape of the apparent unit hydrograph for a given input.

Apparent unit hydrograph

The approach outlined in the last paragraph can also be used where the apparent unit hydrograph is assumed to be represented by a simple conceptual model. In this case the parameters of the model will depend not only on the properties of the catchment, but also on the characteristics of the particular input for a given event. The approach can be illustrated readily for the case where it is assumed that the input shape remains constant from event to event. The only variation is in the mean value of the input x. We then seek the apparent unit hydrograph for the ith input–output event by solving the equation

$$y_i(t) = h_i(t) * \bar{x}_i f(t) \tag{8.34}$$

where $y_i(t)$ is the output for the ith event, $f(t)$ is the constant input shape, $\bar{x}_i$ is the mean rate of input for the ith event, and $h_i(t)$ is the apparent unit hydrograph derived by treating the data, as if the input were transformed to the out in this event, in a linear time-invariant fashion.

If the analysis is done on the basis of a three-parameter conceptual model, we can express the apparent unit hydrograph $h_i(t)$ as a function of time and of the three parameter values which give the closest fit in equation (8.34) for the particular event i.e.

$$h_i(t) = \phi(t, a_i, b_i, c_i) \tag{8.35}$$

The next step in the linearisation approach is to seek the relationship between each of the parameters a, b, c and the level of input x. If we can obtain such a relationship, it is possible to predict the values of the parameters a, b, c for any given level of x and hence to determine the apparent unit hydrograph $h(t)$ for this level of input. If this has been successfully accomplished, the apparent unit hydrograph corresponding to the design level of input can be convoluted with that input to give a prediction of the output based on the linearisation of the system and the correlation of the response parameters with levels of input.

Linearisation can also be applied to the basic non-linear equations of physical hydrology. Solutions of these linearised equations can be used to study the general behaviour of systems but have the disadvantage that certain phenomena, which occur in non-linear systems, do not appear in their linearised versions. The linearisation of the Richards equation for unsaturated flow of soil moisture and the linearisation of the Boussinesq equation for groundwater flow have already been discussed in Section 7.6 above. Accordingly, attention will be concentrated here on the linearisation of the equations for open channel flow.

Linearisation of the equations for open channel flow

The general non-linear equation for unsteady free surface flow in a wide rectangular channel was mentioned in Section 8.1 above. The equation of continuity for the case of routing an upstream inflow (i.e. $r = 0$) is given by

$$\frac{\partial q}{\partial x} + \frac{\partial y}{\partial t} = r(x, t) \tag{8.3}$$

and equation for the conservation of linear momentum for Chezy friction is given by

$$(gy^3 - q^2)\frac{\partial y}{\partial x} + 2qy\frac{\partial q}{\partial x} + y^2\frac{\partial q}{\partial t} = S_0 gy^3 - \frac{g}{C^2}q^2 \tag{8.4b}$$

This highly non-linear equation can be linearised (Dooge and Harley, 1967) by considering a perturbation about a steady uniform flow q_0 and the following linear equation derived for the perturbed discharge

$$(1 - F_0^2)\, gy_0 \frac{\partial^2 q}{\partial x^2} - 2u_0 \frac{\partial^2 q}{\partial x \partial t} - \frac{\partial^2 q}{\partial t^2} = 3gS_0 \frac{\partial q}{\partial x} + \frac{2gS_0}{u_0}\frac{\partial q}{\partial t} \tag{8.36}$$

where the coefficients depend on the assumed reference discharge q_0 about which the perturbation is taken but do not depend on the variable discharge q. Strictly speaking equation (8.36) is only valid for perturbations small enough that the variations of the coefficients in the non-linear equation are not sufficient to produce large errors in the result. In fact, we may consider (8.36) as a version of the non-linear equation in which the coefficients are frozen at the value corresponding to the reference discharge. The use of equation (8.36) for large perturbations is equivalent to the acceptance of the unit hydrograph approach, which assumes that the runoff process remains linear at all levels of flow (see Appendix D).

Linear channel response

Since equation (8.36) is linear, it is only necessary to determine the solution for a delta function input, since the outflow of the downstream end for any other inflow can be obtained by convolution. The impulse response of a channel obtained in this way can be referred to as the linear channel response (LCR) of the channel reach in question (Dooge and Harley, 1967). Any of the standard mathematical techniques for the solution of linear partial differential equations can be used in order to find the solution to equation (8.36); but the Laplace transform method is probably the most convenient. When the latter method is used (Dooge, 1967), the system function, i.e. the Laplace transform of the impulse response (or LCR), is found to be (see Appendix D)

$$H(x,s) = \exp(\lambda_2(s) \cdot x) \tag{8.37a}$$

where

$$\lambda_2 = e \cdot s + f - \sqrt{a \cdot s^2 + b \cdot s + c} \tag{8.37b}$$

and a, b, c, e and f are parameters which depend on the hydraulic characteristics of the channel. Since the above system function is of exponential form, the cumulants of the response function can be determined by repeated differentiation of the quantity inside the brackets in the equation and evaluated at $s = 0$ (see Chapter 3.5, page 51). When this is done and the values of the parameters are substituted in the result, we obtain the first three cumulants of the linear channel response as follows

$$k_1 = U_1' = \frac{x}{1.5u_0} \tag{8.38a}$$

$$k_2 = U_2 = \frac{2}{3}\left(1 - \frac{F^2}{4}\right)\left[\frac{y_0}{S_0 x}\right]\left[\frac{x}{1.5u_0}\right]^2 \tag{8.38b}$$

$$k_3 = U_3 = \frac{4}{3}\left(1 - \frac{F^2}{4}\right)\left(1 + \frac{F^2}{2}\right)\left[\frac{y_0}{S_0 x}\right]^2\left[\frac{x}{1.5u_0}\right]^3 \tag{8.38c}$$

Dimensionless cumulants   The higher cumulants can be made dimensionless by dividing by the appropriate power of the lag and the resulting dimensionless cumulants or shape factors are seen to be functions of the Froude number and the dimensionless length parameter D defined by

$$D = \frac{S_0 x}{y_0} \tag{8.39}$$

Consequently, even if we were unable to invert the transform function given by equation (8.10), it would be possible to determine the cumulants of the solution and to plot the solution of the linearised equation (8.36) on a shape factor diagram. It is clear from equation (8.38) above, that there is a general relationship between the shape factor s_R and the dimensionless length D of the form

$$s_R = \frac{\Phi(F)}{D^{R-1}} \tag{8.40}$$

where $\Phi(F)$ depends only on the Froude number. In particular, for a vanishingly small Froude number we have

$$s_2 = \frac{2}{3D} \tag{8.41a}$$

$$s_3 = \frac{4}{3D^2} \tag{8.41b}$$

so the solution for a vanishingly small Froude number would plot on (s_3, s_2) shape-factor diagram as the line

$$s_3 = 3 s_2^2 \tag{8.42}$$

Further information about the behaviour of the solution can be obtained from an examination of the results for the first moment about the origin given by equation (8.38a) and for the second moment about the centre given by equation (8.38b). The lag given by equation (8.38a) indicates that, for the linearised solution, the average rate of propagation of the flood wave is 1.5 times the velocity corresponding to the reference discharge. This corresponds to the value for the celerity c of the flood wave in a wide rectangular channel with Chezy friction given by

Celerity c of the flood wave

$$c = \frac{\partial Q}{\partial A} = \frac{\partial q}{\partial y} = \frac{3}{2} C S_0^{1/2} y^{1/2} = 1.5u \tag{8.43}$$

which was first proposed by Kleitz (1877) and Seddon (1900). The lag given by equation (8.38a) of course depends on the choice of reference discharge q_0 and is seen from that equation to vary inversely with the cube root of the reference discharge. If we examine equation (8.38b) for the effect on the second moment about the centre of the reference discharge we obtain an interesting result. Since the reference discharge refers to uniform flow, we have the following relationships between the reference discharge q_0, the reference velocity u_0, and the reference depth y_0:

$$q_0 = u_0 y_0 \tag{8.44a}$$

$$u_0 = C S_0^{1/2} y_0^{1/2} \tag{8.44b}$$

$$q_0 = C S_0^{1/2} y_0^{3/2} \tag{8.44c}$$

When the reference discharge is altered, the variation of y_0 in the numerator of equation (8.38b) is compensated for by the variation of u_0^2 in the denominator, so that the second moment about the centre, which measures the dispersion of the linear channel response, is independent of the reference discharge.

Actually, it is possible to invert the system function given by equation (8.37a) to the time domain, but the result is a complicated one. The solution in the original coordinates (x, t) consists of two terms

$$q(x, t) = q_1(x, t) + q_2(x, t) \tag{8.45}$$

The term q_1 is of the following form

$$q_1(x, t) = \delta\left[t - \frac{x}{c_1}\right] \exp(-px) \tag{8.46}$$

where c_1 is the characteristic wave speed given by

$$c_1 = u_0 + \sqrt{g y_0} \tag{8.47}$$

and where p is a parameter, which depends on channel characteristics, being given by

$$p = \frac{(2 - F_0)}{(F_0 + F_0^2)} \frac{(S_0)}{(2y_0)} \tag{8.48}$$

We see from the above result, that, for the linear solution of the case of a delta function input at the upstream end of a channel, we will have a delta function moving downstream at the head of the wave at the characteristic speed given by equation (8.47). However, its volume declines exponentially at a rate determined by the parameter defined by equation (8.48). It is clear that for conditions resembling those in most natural rivers (i.e. for a Froude number of 0.2 or less) the decline in the volume under the head of the wave given by equation (8.46) will be very rapid.

The second term in equation (8.45) which represents the attenuated body of the wave is given by

$$q_2(x, t) = h \left[\frac{x}{c_1} - \frac{x}{c_2} \right] \exp\left[-rt + sx \right] \frac{I_1[2ha]}{a} \tag{8.49}$$

Characteristic wave speed

where c_2 is the upstream characteristic wave speed given by

$$c_2 = u_0 - \sqrt{gy_0} \tag{8.50}$$

and where $I_1[2ha]$ is a modified Bessel function and a is given by

$$a = \sqrt{[t - x/c_1][t - x/c_2]} \tag{8.51}$$

and r, s and h are parameters depending on the hydraulic properties of the channel (see Appendix D).

The use of the linearised equation is illustrated in Figure 8.9 for a channel with a steady state rating curve given by

$$q_0 = 50 y_0^{3/2} \tag{8.52}$$

and for an upstream inflow between $t = 0$ and $t = 96$ given by

$$q(0, t) = 125 - 75 \cos\left(\frac{\pi t}{48} \right) \tag{8.53}$$

which corresponds to the inflow used by Thomas (1934) in his classical paper on unsteady flow in open channels. Figure 8.9 shows a modification of the flood wave for a distance up to 500 miles.

For any linearisation of the flood routing problem, it is necessary to choose a value of the reference discharge q_0 about which the discharge is perturbed. Since this value q_0 is used to evaluate the reference depth y_0 from equation (8.44a) and hence the reference velocity u_0 from equation (8.44b) and finally determines the coefficients in the linearised equation (8.36), the choice of the reference discharge will naturally affect the result.

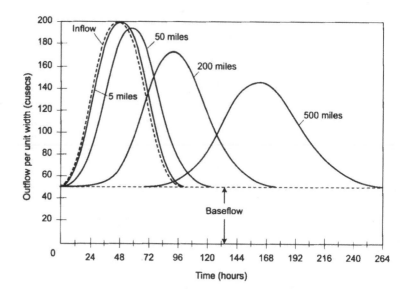

Figure 8.9. Linear flood routing (Dooge and Harley 1967).

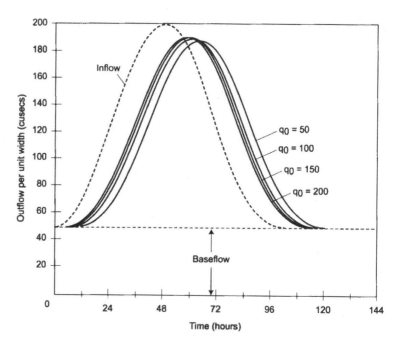

Figure 8.10. Effect of reference discharge (Dooge and Harley 1967).

Figure 8.10 shows the effect of reference discharge on the outflow 50 miles downstream for the channel characterised by equation (8.52) and the inflow given by equation (8.53). In the latter equation the inflow varies from 50 cubic feet per second per foot width to 200 cubic feet per second per foot width. Accordingly the reference discharge was taken at values of

Flood waves

50, 100, 150 and 200 cubic feet per second per foot width. It can be seen from Figure 8.10 that the resulting flood waves are displaced in time, as might be expected, since the lag given by equation (8.38a) depends on the reference velocity u_0 and hence on the reference discharge q_0. However, it is interesting to note that the shape of the flood wave and the value of the peak are very similar for the various reference discharges. This would appear to be a reflection of the fact that the second moment about the centre, which measures the dispersion, is given by equation (8.38b) and is independent of the reference discharge. It would appear to indicate that the effect of the third and higher cumulants, which are affected by the reference discharge, have little influence on the shape of the outflow. It also suggests that the use of a linear model for flood routing might be expected to show greater accuracy in prediction of the peak outflow than in regard to the time to peak.

Subsequent work on this problem between 1967 and 1989 has been well reviewed by Napiorkowski (1992). His account covers the work done on such topics as (a) the generalisation of the approach from a wide rectangular channel with Chezy friction to the general case of any channel shape and any friction law; (b) the extension from a semi-infinite channel with an upstream condition to a finite channel with both upstream and downstream control; (c) general expressions for the moments and cumulants of any order of the linearised channel response; (d) the attenuation and phase shift for a harmonic input to a linearised channel.

Friction law

The general form of the linearised equations for the case where the friction law takes the form $Q = kA^m$ can be derived easily. Such an assumption is almost always adopted in the study of both steady and unsteady flow in open channels. For Chezy friction the value of m varies from $m = 3/4$ for a triangular cross section to $m = 3/2$ for a wide rectangular channel. For Manning friction the corresponding values are $m = 5/6$ for a triangular cross section and $m = 5/3$ for a wide rectangular channel. The parameter m corresponds to the ratio of the kinematic wave velocity (U_k) to the average velocity of flow ($\bar{u}$). When this generalisation is made, the linearised equation becomes

$$(1 - F_0^2)\bar{y}_0 \frac{\partial^2 Q}{\partial x^2} - 2u_0 \frac{\partial^2 Q}{\partial x \partial t} - \frac{\partial^2 Q}{\partial t^2} = 2mgS_0 \frac{\partial Q}{\partial x} + \frac{2gS_0}{u_0} \frac{\partial Q}{\partial t} \qquad (8.54)$$

where $\bar{y}_0$ is the value at the reference condition of the hydraulic mean depth, i.e. the ratio of the area of flow to the surface width and u_0 is the velocity corresponding to the reference conditions used as a basis for linearisation. The cumulants given by equation (8.38) above are then generalised (Dooge, Napiorkowski and Strupczewski, 1987b) to

$$k_1 = \frac{x}{mu_0} \qquad (8.55a)$$

$$k_2 = \frac{1}{m}\left(1 - (m-1)^2 F_0^2\right)\left[\frac{y_0}{S_0 x}\right]\left[\frac{x}{mu_0}\right]^2 \tag{8.55a}$$

$$k_3 = \frac{3}{m^2}\left(1 - (m-1)^2 F_0^2\right)\left(1 + (m-1)F_0^2\right)\left[\frac{y_0}{S_0 x}\right]^2\left[\frac{x}{mu_0}\right]^3 \tag{8.55b}$$

In general, it can be shown that all higher cumulants can be expressed in the form

$$k_R = \phi(m, F_0)\left[\frac{y_0}{S_0 x}\right]^{R-1}\left[\frac{x}{mu_0}\right]^R \tag{8.56}$$

Closed form expressions can be developed for the function $\phi(m, F_0)$ for any order of cumulant (Romanowicz, Dooge and Kundzewicz, 1988).

Flood routing   The above results relate to the classical problem of flood routing, i.e. downstream movement of water with a given upstream control and negligible effect of the downstream control, i.e. a semi-infinite channel. The relative effects of the upstream and controls for a finite channel reach were first solved for the case of the diffusion analogy which corresponds to the assumption that the Froude (F_0) number is vanishingly small (Dooge and Napiorkowski, 1984). The solution for the case of any value of m and any value of F_0 was derived by the same authors (Dooge and Napiorkowski, 1987). The downstream and upstream waves are each reflected at the other boundary and consequently the two single terms representing the transformation of the upstream boundary condition and the upstream transformation of the downstream boundary condition respectively are replaced by two corresponding infinite series of similar form with strong convergence. The rate of convergence is the same for these infinite series and is given by

$$\alpha = \exp\left[\frac{-2mS_0 L}{y_0(1 - F_0^2)}\right] \tag{8.57}$$

so that the convergence is very rapid except for small values of y_0 and F_0. In practical usage, a close approximation is obtained by including only the first term in each of the series corresponding to (a) the initial downstream propagation of the upstream boundary condition, (b) the initial upstream propagation of the downstream boundary condition, and (c) the first reflection of the downstream propagation at the downstream boundary (Napiorkowski and Dooge, 1988).

Sinusoids   An extension can be made from the analysis of the response to a delta function input to the more general case, by seeking the response to an upstream inflow of an infinite periodic train of *sinusoids*, and hence the response to any function which can be broken down into a series of such functions. (Kundzewicz and Dooge, 1989). On the basis of the continuity equation alone, it is possible to establish that the necessary and sufficient condition for the attenuation of a flood wave is identical to the necessary and sufficient condition for the existence of a phase shift between the flow rate and the area of flow and for the existence of a looped rating curve. The common condition is that either the frequency ω or the wave length β is complex. If the upstream inflow is taken as a train of non-attenuating waves, then the frequency must be real. If the wave number is also real then there can be no attenuation with distance and we have the limiting case of kinematic wave propagation.

If the wave number is taken as complex then we have

$$f(x, t) = f_0 \exp(-\beta_I x) \cos(\omega_R t - \beta_R x) \tag{8.58}$$

where ω_R is the real frequency, β_R is the real part of the wave number, and β_I is the imaginary part of the wave number. It can be shown for the above case of $\omega_I = 0$ that the logarithmic decrement δ, i.e. the logarithm of the attenuation over a single cycle, is related to the phase shift ϕ by the equation

$$\delta = \tan \phi \tag{8.59}$$

For $F_0 = 0$ (i.e. the diffusion analogy) attenuation over a single cycle varies from unity (no decrease) for very small frequencies to zero (filtering out) for very high frequencies and the phase shift varies from 0° for very small frequencies to 45° for very high frequencies. For the case of the kinematic wave i.e.

$$F_0 = \frac{1}{m - 1} \tag{8.60}$$

the attenuation factor is unity (i.e. no reduction) and the phase shift is zero for all frequencies between zero and infinity. For intermediate values of F_0 and for frequencies in the range

$$10^{-2} < \frac{y_0}{S_0 U_0} \omega_R < 10^2 \tag{8.61}$$

the attenuation factor drops sharply from unity to zero and the phase shift increases from zero to a maximum value and then declines again to zero. Figure 8.11 (Kundzewicz and Dooge, 1989) shows this variation in term of the two dimensionless variables. Dimensionless frequency ω_R' and dimensionless attenuation factor β_1' (Kundzewicz and Dooge, 1989).

The above linearisation approach can also be applied to the case of a uniform channel with uniform lateral inflow and no upstream inflow. The solution (O'Meara, 1968) is more complex than the case of upstream inflow without lateral inflow dealt with above. Two facts are noteworty. First the s_2–s_3 relationship for the case of lateral inflow shows less dependence on the Froude number of the reference flow than the spread for the

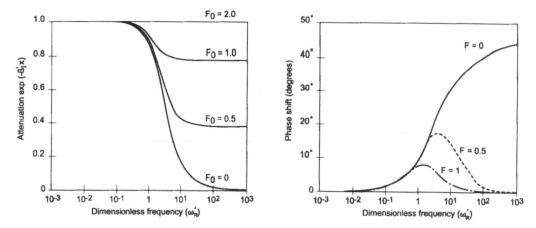

Figure 8.11. Attenuation and phase shift in LCR (Kundzewicz and Dooge, 1989).

case of upstream inflow where the shape factors are based on the values of the cumulants given by equation (8.38) above. Secondly, the shape factor relationship shows a close relationship with that obtained for the case of uniform lateral inflow to a cascade of linear reservoirs given by equations (5.48) to (5.50) above.

8.4 NON-LINEAR BLACK-BOX ANALYSIS

In seeking an algebraic relationship between a dependent variable y and an independent variable x, we first examine whether the convenient linear relationship would be adequate for our purpose and if we are not satisfied with this simple regression model generalise the relationship to a polynomial one. In the same way the linear system relationship can be generalised to a polynomial relationship. The first-order convolution relationship represented by

$$y(t) = \int_0^t h(\tau)x(t - \tau)\,d\tau \tag{8.62}$$

can be generalised to an nth order system in which the input $x(t)$ and the output $y(t)$ are connected by the relationship

$$y(t) = \int_0^t \int_0^t \cdots \int_0^t h_n(\tau_1, \tau_2, \ldots, \tau_n) \prod_{i=1}^{n} x(t - \tau_i)\,d\tau_i \tag{8.63}$$

and the system operation is described by the nth order kernel function $h_n(t_1, t_2, \ldots, t_n)$. A general non-linear functional representation of the relationship between input and output can be obtained by forming a series of terms of the type of equation (8.63) thus obtaining what is usually termed a Volterra series:

$$y(t) = \sum_{n=1}^{N} \int_0^t \cdots \int_0^t h_n(\tau_1, \ldots, \tau_n) \prod_{i=1}^{n} x(t - \tau_i) \, d\tau_i \qquad (8.64)$$

The first term of the above series is the linear convolution relationship of equation (8.64).

The representation given by equation (8.64) was introduced by Wiener (1942, 1958) in communication engineering and was first applied to hydrologic systems by Amorocho and Orlob (1961). Further applications of the approach to the hydrologic systems are to be found in papers by Amorocho (1963), Jacoby (1966), Harder and Zand (1969), Amorocho and Brandstetter (1971), Hino, Sukigara and Kikkawa (1971), Kuchment and Borshevsky (1971), Diskin and Boneh (1972). If the data for the input and the output are only available at discrete intervals then the general non-linear relationship of equation (8.64) is written in the analogous discrete form as

$$y(s) = \sum_{n=1}^{N} \sum_{\sigma_1=0}^{s} \cdots \sum_{\sigma_n=0}^{s} h_n(\sigma_1, \sigma_2, \ldots, \sigma_n) \prod_{i=1}^{n} (s - \sigma_i) \qquad (8.65)$$

As in the case of the simple first-order convolution relationship, the multi-dimensional convolution represented by equation (8.65) can be expressed in matrix form as

$$\underline{y} = \underline{X}\,\underline{h} \qquad (8.66)$$

but the simplicity of form of the latter equation conceals much of the complexity of the problem. The output vector y corresponds as before to the known values of the output at discrete intervals of time. In this case, however, the vector elements of unknown ordinates must contain sufficient values to define all of the n-dimensional response functions. In addition, the elements that go to make up the matrix X consist not only of the values of the input vector x but all possible products of these values.

As in the case of the special case of first order convolution, which applies to linear time-invariant systems, there are many possible approaches to the problem of identifying the non-linear system functions. As might be expected our experience in regard to linear systems (described in Chapter 4) is of considerable help in this connection. It is of course possible to attempt to solve equation (8.65) by direct inversion, or by means of least squares optimisation. However, the size of the matrix to be inverted creates special difficulties. If it is decided that $(r + 1)$ ordinates are needed to represent a response function in any one dimension, then an n-dimensional response function will be defined by $(r + 1)^n$ elements.

However, all system functions of second order or higher are symmetric and consequently the number of independent values to be determined is reduced from $(r + 1)^n$ to the value given by

$$R_n = \binom{r+n}{n} \tag{8.67}$$

where R_n is the total number of independent ordinates in an n-dimensional system function, each dimension contains $(r + 1)$ elements. Consequently if we wish to analyse a non-linear system by equation (8.65) the total number of ordinates to be determined is given by

$$R(N) = \sum_{n=1}^{N} \binom{r+n}{n} \tag{8.68}$$

where N is the highest order term contained in the functional expansion.

Even for relatively low values of r and N, the number of ordinates to be determined can become quite large. Thus, if the sampling interval is one-tenth of the memory length of the system (this is probably the upper limit for any reasonably degree of resolution), the number of ordinates to be determined in accordance with equation (8.66) will be 11 for the first-order system function, 66 for the second-order system function, 286 for the third-order system function, and 1,001 for the fourth-order system function. It is clear that the order of the matrix to be inverted increases rapidly and would soon become impossible to handle even on a large computer. A further difficulty is that the number of ordinates to be determined may exceed the length of the record of runoff, which would consequently not contain enough information for the determination of all the ordinates even if the output data were perfect. Consequently, the approach based on direct matrix inversion, or on the use of least squares, runs into considerable trouble quite apart from the problems of the effect of errors on numerical stability which were illustrated for the linear case in Chapter 4.

In the case of linear black-box analysis, it was seen that one of the most successful methods of identification was expansion in terms of orthogonal functions. Harder and Zand (1969), and Amorocho and Brandstetter (1971) have used orthogonal functions in the non-linear case. This will *Third-order system* be illustrated below for the case of a pure third-order system function, *function* but the approach can be used for any order of function and for a series of system functions as represented by equation (8.64) above. Since a third-order system response can be considered as the product of three first-order responses, the expansion of each of these first-order responses in terms of a given set of orthogonal functions will be equivalent to the expansion of a third-order system response in the terms of triple products of the orthogonal functions. Thus, in the case of third-order system functions, we could express its value for any set of co-ordinates (s_1, s_2, s_3) as

$$h_3(s_1, s_2, s_3) = \sum_i^{M_3} \sum_j^{M_3} \sum_k^{M_3} \alpha_{ijk} F_i(s_1) F_j(s_2) F_k(s_3) \tag{8.69}$$

where M_3 is the number of terms used in the expansion (taken as equal in all three dimensions) and $F(s)$ belongs to a family of complete orthogonal functions. Each of the orthogonal functions used to construct the system response given by equation (8.69) can be convoluted with the input in order to compute the function

$$a_i(s) = \sum_{\sigma=0}^{s} F_i(\sigma) x(s - \sigma) \tag{8.70}$$

For the case of a pure third-order system equation (8.65) will take the form

$$y_3(s) = \sum_{\sigma_1=0}^{s} \sum_{\sigma_2=0}^{s} \sum_{\sigma_3=0}^{s} h_3(\sigma_1, \sigma_2, \sigma_3) \prod_{i=1}^{3} (s - \sigma_i) \tag{8.71}$$

Substituting from equation (8.69) into equation (8.71) interchanging the order of the summations, separating the sums with respect to the arguments σ_1, σ_2 and σ_3 and using the definition of $a_i(s)$ given by equation (8.70) we obtain

$$y_3(s) = \sum_{i=0}^{M_3} \sum_{j=0}^{M_3} \sum_{k=0}^{M_3} \alpha_{ijk} a_i(s) a_j(s) a_k(s) \tag{8.72}$$

Separating out the terms, where either two values or three values of the indices (i, j, k) are equal, and using the result that the coefficients of the expansion of a symmetric kernel are also symmetric (Brandstetter and Amorocho, 1970) we obtain

$$y_3(s) = \sum_{k=0}^{M_3} \alpha_{kkk} a_k(s) a_k(s) a_k(s)$$

$$+ 3 \sum_{j=1}^{M_3} \sum_{k=0}^{j+1} \alpha_{jkk} a_j(s) a_k(s) a_k(s)$$

$$+ 6 \sum_{i=2}^{M_3} \sum_{j=1}^{i+1} \sum_{k=0}^{j+1} \alpha_{ijk} a_i(s) a_j(s) a_k(s) \tag{8.73}$$

Despite its appearance, equation (8.73) represents a great simplification as compared with equation (8.71), since the values of the functions such as $a_i(s)$ can be found from the known input once the functions are chosen to serve as a basis for expansion. The only unknowns in equation (8.73) are the coefficients α_{ijk} and the equation is seen to be linear in these unknown coefficients. Accordingly, we see that, by use of expansion in terms of orthogonal functions, the question has been transformed from a non-linear problem to a linear problem. In order to obtain reasonable values for these unknown coefficients, it is of course necessary that the number of output data points should exceed the number of unknown coefficients. The number of unknown coefficients will be given by the same formula as in equation (8.67) above, but in this case r represents the highest order of function $F_r(s)$ used in the expansions. Amorocho and Brandstetter (1971) have used Meixner functions for the expansion

Non-linear kernels

of non-linear kernels and found that between five and eight terms were sufficient for a good approximation. This compares with between ten and twenty ordinates used for the definition of a kernel in each dimension in the time domain. Examples of the use of Meixner functions in non-linear black-box analysis, for both synthetic data and for field data, are described by Amorocho and Brandstetter (1971).

The pioneering work of Amorocho, Orlob and Brandstetter at the University of California in Davis and of Diskin and Boneh at the Technion in Haifa has since been supplemented by the work of Napiorkowski (1983, 1984 and 1986) at the Institute of Geophysics in Warsaw. A notable element in his contribution was the derivation of a method to simplify the derivation of the higher-order derivates once the second-order system function has been obtained thus by-passing the difficulty of using data to evaluate the huge number of ordinates required according to equation (8.68). Another element of his work was the reduction of the general case described above to a 4-parameter model consisting of a pure translation and a cascade of equal non-linear reservoirs (Napiorkowski and O'Kane, 1984). This quadratic approximation is found to give acceptable accuracy. This is an example of uniform non-linearity which is discussed in the next section.

8.5 CONCEPT OF UNIFORM NON-LINEARITY

On the basis or our experience with linear conceptual models, we would expect that some success could be achieved in the non-linear case by the use of simple conceptual non-linear models. As in the linear case we have a wide variety of models to choose from. We have already seen

Non-linear reservoir

in Section 8.2 that the simple conceptual model of a single non-linear reservoir has been applied for a number of years in the solution of overland flow problems. In this case the conceptual model has two parameters (a and c) compared with the single parameter (the storage delay time K) in the case of a single linear reservoir. We would expect non-linear conceptual models to have at least one more parameter than the corresponding linear conceptual models since the degree of non-linearity of each component will itself be an additional parameter. The Nash cascade of equal linear reservoirs is the two parameter conceptual model most widely used in the linear analysis of hydrologic systems. The properties of the special conceptual model consisting of a cascade of equal non-linear reservoirs (Dooge, 1967) which is the non-linear counterpart have been investigated by Dooge (1967). Because all of the reservoirs in the cascade are taken to have the same degree of non-linearity, Dooge referred to such a system as *Uniform non-linearity*   having uniform non-linearity.

If we have a cascade of equal non-linear reservoirs each of which behaves in accordance with the relationship

$$Q(t) = a[S(t)]^c \tag{8.74}$$

where $Q(t)$ is the outflow from the reservoir, $S(t)$ is the storage within the reservoir and a and c are parameters characterising the reservoir. If the reservoirs form a cascade then the outflow from each reservoir constitutes the inflow to the reservoir immediately downstream from it. Accordingly for the ith reservoir in the cascade the governing equation will be

$$Q_{i-1}(t) + r_i(t) - Q_i(t) = \frac{dS_i}{dt} \tag{8.75}$$

where Q_{i-1}, is the discharge from the upstream reservoir, r_i is the lateral inflow into the ith reservoir, Q_i is the discharge from the ith reservoir, and S_i is the storage in the ith reservoir. Substituting from equation (8.74) and rearranging gives us

$$\frac{dS_i}{dt} + aS_i^c = aS_{i-1}^c + r_i \tag{8.76}$$

where S_i, S_{i-1}, and r_i are all functions of time. To predict the outflow from a system characterised by equation (8.76) for a given set of inflows $(r_0, r_1, r_2, \ldots, r_n)$ it is necessary to solve the equations successively for each of the n reservoirs in the cascade. For the special case of $c = 2$ equation (8.76) has the form of a Riccati equation and can be linearised by a special transformation. Even for this case, however, it is difficult to obtain an analytical solution even for a constant inflow for a cascade of a small number of reservoirs. Rather than attack the problem in this direct and laborious fashion, we will investigate the behaviour of a system governed by equations of the type given by equation (8.76) and see if it has any special properties that might be of use in hydrologic analysis.

In examining the properties of any set of equations it is advisable to express them in dimensionless form. This can be conveniently done if we choose a value of $\bar{r}$ which characterises in some way the level of intensity of input and derive from this the equivalent storage in accordance with equation (8.74) thus obtaining

$$\bar{r} = a(\bar{S})^c \tag{8.77a}$$

or its equivalent

$$\bar{S} = \left(\frac{\bar{r}}{a}\right)^{1/c} \tag{8.77b}$$

We can now define the inflow into each reservoir in dimensionless form, as

$$r'_i = r_i/\bar{r} \tag{8.78}$$

and the storage in each reservoir in dimensionless form as

$$S'_i = S_i/\bar{S} \tag{8.79}$$

It remains to define a characteristic time, which can be used as the basis of the dimensionless time to be used in our dimensionless equation. Since we have already defined the characteristic storage $(\bar{S})$ which has the dimension of volume and a dimensionless discharge $(\bar{r})$ which has the dimension of volume per unit time, it is obviously convenient to use these two to define a characteristic time as

$$\bar{t} = \frac{\bar{S}}{\bar{r}} = \frac{1}{(a)^{1/c}(\bar{r})^{(c-1)/c}} \tag{8.80}$$

and consequently the dimensionless time variable as

$$t' = t/\bar{t} \tag{8.81}$$

It is important to note that, because of the non-linearity inherent in the relationship between $\bar{r}$ and $\bar{S}$ as expressed by equation (8.77a), the value of the characteristic time $\bar{t}$ will vary with the characteristic rate of inflow $\bar{r}$.

Since the characteristic discharge $(\bar{r})$, the characteristic storage $(\bar{S})$ and the characteristic time have been defined in this way, we have the identity

$$\bar{S}/\bar{t} = \bar{r} = a(\bar{S})^c \tag{8.82}$$

and we can divide each term in equation (8.76) by one or other of the equal quantities in (8.81) to obtain the dimensionless form of the equation as

$$\frac{dS'_i}{dt'} + (S'_i)^c = (S'_{i-1})^c + r'_i(t) \tag{8.83}$$

For any fixed value of c and for the fixed pattern of inputs denoted by the dimensionless input vector $\underline{r}'$, the solution of the set of equations represented by equation (8.83) will be of the general form

$$S'_i(t') = f[t', c, \underline{r}'(t)] \tag{8.84}$$

even if we are unable to determine the nature of the unknown function $f[\,]$. Similarly, since the discharge from the total system will be related to the storage in the final reservoir by

$$Q'_n(t') = (S'_n)^c \tag{8.85}$$

we know that the dimensionless discharge will be represented by a function of the general form

$$Q'_n(t') = g[t', c, \underline{r}'(t')] \tag{8.86}$$

where r' is the dimensionless input for a single event. The latter equation is used as the basis of discussion of the special properties of a system consisting of a cascade of equal non-linear reservoirs.

A key feature of the behaviour of non-linear systems (and one of great importance in hydrology) is the effect on the behaviour of the system, of *Level of intensity of the* change in the level of intensity of the input. If we have two events, in which *input* the input pattern to the cascade of equal non-linear reservoirs is the same, except that the intensities in one are a constant ratio of those in the other, then (provided the index of non-linearity c is greater than one) the higher intensity event will result in more rapid runoff and consequently a different pattern of output. Our concern here is the effect of such a difference of level in intensity on the solution of the set of equations represented by equation (8.83). When the level of intensity of input $r(t)$ changes but the pattern remains the same, the value of $\underline{r}'(t)$ will not change because all intensities including the mean intensity (or other characteristic values) $\bar{r}$ will change in the same ratio. The effect on the solution of each of the equations in turn derives from the fact that, since the characteristic time $\bar{t}$ depends on the level of intensity, it will be different in the two events. Consequently, two inflows which are the same, when expressed as $r'(t)$, will show two different patterns of input, when each is expressed as $r'(t')$, since any time characteristic, which was equal in the original time domain, will become unequal in the dimensionless time domain due to the two different values of $\bar{t}$. This difficulty will not arise, if the pattern of inflow does not contain any characteristic times such as, for example, where the input of the system consists of a step function input distributed in a fixed ratio among the reservoirs in the cascade. In the latter case, r does not depend on time. Consequently, the dimensionless outflow Q'_n, will be a function only of t' for any given value of c and any fixed distribution of input among the reservoirs. This solution will indeed be the dimensionless *Rising hydrograph* rising hydrograph for this particular type of cascade and distribution of inflow.

The considerations discussed in the last paragraph lead to the concept of certain input patterns being similar in the uniform non-linear sense.

Thus if we have two vectors of inflows for a cascade, $\underline{r}_1(t)$ and $\underline{r}_2(t)$, such that

$$\underline{r}_1'(t) = \underline{r}_2'(t) \tag{8.87}$$

these inputs could be defined as being similar in the sense of uniform non-linearity. In order for such similarity to exist, it would not be necessary that the general pattern of the input should be the same, but that all of the time characteristics of the individual events should be related to the level of intensity of the input in accordance with equation (8.72). Thus for two input events to be similar, the one with the higher intensity would have to have shorter time characteristics and shorter by such an extent that equation (8.79) would be satisfied. In this case equation (8.86) can be written as

$$Q_n'(t') = g[t', c, \underline{r}'(t')] \tag{8.88}$$

where r' applies to all events with similar dimensionless inputs so that for a fixed value of c and similar inputs the dimensionless output form the system would be a function only of the dimensionless time.

Fixed pattern of input   Consequently for any fixed pattern of input, we could determine the shape of the dimensionless response defined by equation (8.83) from a single input-output event and use it together with equation (8.80) to predict the response for any event which was similar in the sense of uniform non-linearity. For a step function input, there is no characteristic time and thus we would have a single dimensionless response. For the case of constant uniform inflow for a period D, we would have

$$\frac{Q_n(t)}{r} = \Phi\left[\frac{t}{\bar{t}}, \frac{D}{\bar{t}}\right] \tag{8.89}$$

where the functional form would depend on the value of c.

Both the Horton–Izzard solution and the kinematic wave solution to the problem of overland flow discussed in Section 8.2 can be expressed in the form of equation (8.89). In the case of the Horton–Izzard solution, this is to be expected, since in this case we are dealing with a single non-linear reservoir, which is a special case of the cascade discussed above. The kinematic wave solution can also be considered as a limiting case of uniform non-linearity. It will be recalled from Chapter 5, that the limiting case of a cascade of equal linear reservoirs, in which the number of reservoirs n tended to infinity while the storage delay time of each reservoir K tended to zero in such a way that the lag of the system nK remained finite, was seen to be equivalent to a linear channel with a lag time equal to nK and this to represent a pure translation. In the case of a cascade of equal non-linear reservoirs a corresponding relationship exists. It can be shown that the kinematic wave solution corresponds to the limiting case of an infinitely long cascade of infinitesimal non-linear reservoirs. This suggests that the two simple approaches to overland flow discussed in the Section 8.2, represent limiting cases of the conceptual model based on uniform non-linearity, and that solutions intermediate between the two, would be obtained by the simulation of overland flow by a cascade of equal non-linear reservoirs of finite length.

There remains the question of whether the non-linear conceptual model discussed above is of any use in simulating the observed non-linear behaviour of hydrologic systems. In Section 8.1 the results obtained from a small test basin by Amorocho and Orlob (1961) were mentioned as an example of careful laboratory measurements which demonstrated non-linear behaviour on a hydrologic system of laboratory scale (see Figures 8.2 and 8.3 above). It is interesting to test whether the concept of uniform non-linearity would be adequate to simulate the non-linearity shown by these data. The results for a step function were given in terms of outflow volume versus time. Because of some uncertainty as to whether the tests were carried to full equilibrium and because of the cumulative nature of the data, the characteristic time was defined as being the time at which cumulative outflow was equal to one half of the cumulative inflow. For any steady inflow of intensity q_e this characteristic time t_0 is defined by

$$\int_0^{t_0} q(t)\, dt = \frac{q_e t_0}{2} \tag{8.90}$$

Step function input

and t_0 can be found from the recorded values of cumulative outflow and the intensity of inflow q_e. For the case of a step function input, the approach based on uniform non-linearity suggests that there should be a single relationship of the form

$$\frac{q}{q_e} = f\left[\frac{t}{t_0}\right] \tag{8.91a}$$

or in cumulative form

$$\frac{\sum q}{q_e t_0} = f\left[\frac{t}{t_0}\right] \tag{8.91b}$$

Figure 8.12 shows a plotting of the data of Amorocho and Orlob (1961) in this form. It can be seen that for the three levels of intensity all the points plot along a single undefined curve, thus indicating that this particular non-linear hydrologic system shows the behaviour of a uniform non-linear system. The characteristic time t_0 was found to be related to the rate of inflow (or equilibrium rate of outflow) q_e by

$$t_0 = \frac{281.4}{q_e^{0.3997}} \tag{8.92}$$

Rounding the exponent to 0.4 indicates that the value of the index of non-linearity is $c = 5/3$ which corresponds to rough turbulent flow. It would be reasonable to predict the cumulative outflow for any other intensity of input to this basin on the basis of the curve shown in Figure 8.12 and

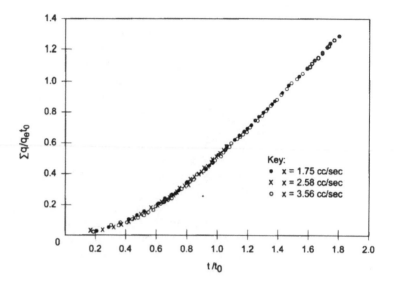

Figure 8.12. Dimensionless
plot of laboratory data.

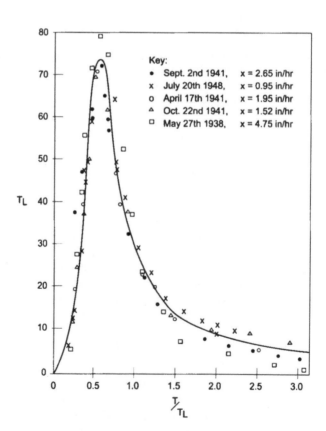

Figure 8.13. Dimensionless
plot of field data.

to convert from dimensionless to actual terms by the use of the relationship given by equation (8.89).

The unit hydrograph method depends on the assumption that the catchment acts as a linear system and hence that the unit hydrograph would be invariant for all shapes and intensity of inflow. Figure 8.1 in Section 8.1 shows the "unit hydrographs" derived by Minshall (1960) for a small catchment of 27 acres and clearly indicates that for the range of intensities shown there is a three-fold variation in the peak of the unit hydrograph. These "unit hydrographs" were re-plotted in dimensionless form by taking the lag as the characteristic time. According to the hypotheses of uniform non-linearity a single dimensionless response curve would be obtained if the ratio of duration to lag were the same in each storm. In fact no knowledge of the duration is available from the published paper so that the data may include quite a variation in the ratio of duration to lag. The result of this dimensionless plotting is shown on Figure 8.13.

Shapes and intensity of inflow

It will be seen from the figure that the points seem to cluster around a single curve and that the variations from this single curve are very much less than those in the unit hydrograph (i.e. the apparent linear responses) shown in Figure 8.1. In this case also a single curve could be used for prediction purposes and the scale of the curve determined by relating the lag to the intensity of effective precipitation.

The above results are sufficiently encouraging to suggest that the conceptual model based on a cascade of equal non-linear reservoirs should be a flexible and reasonably satisfactory tool in the handling of non-linear problems in hydrology. Singh (1988) has applied the model with an inflow into the upstream reservoir only, to field data and discussed the question of the estimation of the parameters c, a and n.

Problem Set

RUNOFF PREDICTION

1. The following values

   $0, 2, 4, 3, 2, 1, 0$

   when reduced to unit volume represent the ordinates at hourly intervals of a two-hour unit hydrograph. Carry out the following procedures:
   (a) Determine the runoff if the following values

   $1, 5, 2, 2$

   represent the volume of excess rain in successive two-hour periods.
   (b) Calculate the ordinates of the S-curve and from these, derive the ordinates of the 8-hour unit hydrograph.
   (c) Determine the effect on the predicted maximum runoff of ignoring the variation in the intensity of effective rainfall in the storm given in (a) above.
   (d) Derive the one-hour unit hydrograph.
2. A two-hour unit hydrograph is defined by a triangle of unit volume whose ordinates at hourly intervals are in the proportion:

   $0, 3, 6, 4, 2, 0$

   Carry out the calculations (a), (b), (c) and (d) described in question 1.

SYSTEM IDENTIFICATION

3. If the input to a system is given by

   $2, 6, 1$

   and the output from the system by

   $0, 4, 14, 8, 1, 0$

   find the unit pulse response of the system by
   (a) forward substitution,
   (b) backward substitution,
   (c) the Collins method,

(d) least squares.

(e) the De Laine method (root matching)

Compare the answers and comment.

4. If the output function in problem (3) were mistakenly taken as

$$0, 4, 11, 8, 1, 0$$

What would be the effect on the derived unit hydrograph for each of the methods used in answering question 3 above?

5. Assume that the effective precipitation is given by

$$\frac{1}{2}\left[t + \frac{t^2}{2}\right]\exp\left(-\frac{t}{2}\right)$$

and the hydrograph of storm runoff by

$$\frac{1}{2}\left[\frac{t^5}{5!} + \frac{t^6}{6!}\right]\exp\left(-\frac{t}{2}\right)$$

Determine as accurately as possible the form of the instantaneous unit hydrograph.

UNIT HYDROGRAPH DERIVATION

6. The following table gives the volume of effective rain and the rate of storm runoff for a storm event on the Ashbrook catchment. Find the three-hour unit hydrograph for the catchment.

Data for Ashbrook Catchment (see footnote below table).

Date and time (hours)	Effective rain (Cusecs)	Storm runoff (Cusecs)
Mar-26		
15.00	1829	0
18.00	3530	30
21.00	8330	340
24.00	0	980
Mar-27		
3.00	0	1320
6.00	0	1390
9.00	0	1280
12.00	0	1160
15.00	0	1040
18.00	0	940
21.00	0	790
24.00	0	680
Mar-28		
3.00	0	580
6.00	0	480

(Continued)

Date and time (hours)	Effective rain (Cusecs)	Storm runoff (Cusecs)
9.00	0	390
12.00	0	320
15.00	0	280
18.00	0	240
21.00	0	210
24.00	0	180
Mar-29		
3.00	0	155
6.00	0	135
9.00	0	115
12.00	0	100
15.00	0	85
18.00	0	70
21.00	0	65
24.00	0	60
Mar-30		
3.00	0	55
6.00	0	50
9.00	0	45
12.00	0	40
15.00	0	35
18.00	0	30
21.00	0	25
Mar-31		
24.00	0	15
3.00	0	10
6.00	0	0

1. Cusecs is the non-standard imperial unit of flow, cubic feet per second; 2. The values of storm runoff are instantaneous values sampled on the hour; 3. The corresponding values of effective rain are uniform intensities during the following three-hour period.

7. The following table shows the effective rainfall in inches and the runoff in cubic feet per second for the Big Muddy River at Plumfield in Illinois for April and May 1927 (Sherman, 1932). Derive the 24-hour unit hydrograph from these figures.

Day	Effective rain (inches)	Runoff (Cusecs)
1	0.068	(132)
2	0.000	(176)
3	0.000	(229)
4	0.000	(263)
5	0.037	(312)
6	0.000	(264)

(Continued)

Day	Effective rain (inches)	Runoff (Cusecs)
7	0.027	(167)
8	0.102	(553)
9	0.080	(666)
10	0.440	(1,615)
11	0.083	2,116
12	0.880	4,201
13	0.690	6,188
14	0.820	8,580
15	0.423	10,229
16	0.500	11,481
17	0.000	10,924
18	0.040	9,375
19	0.130	7,355
20	0.033	5,196
21	0.074	3,500
22	0.025	2,377

CONCEPTUAL MODELS

8. Derive by direct integration or convolution the impulse response for each of the following conceptual models:
   (a) a single linear reservoir,
   (b) a cascade of 2 equal linear reservoirs,
   (c) a cascade of 2 unequal linear reservoirs,
   (d) a cascade of 3 equal linear reservoirs,
   (e) a cascade of 4 equal linear reservoirs.
9. Derive the impulse response for each of the following conceptual models:
   (a) a rectangular inflow routed through a linear reservoir,
   (b) a triangular inflow routed through a linear reservoir.
10. Derive the impulse response for each of the following conceptual models:
    (a) a cascade of n equal reservoirs,
    (b) a Muskingum-type reach,
    (c) a convective-diffusion process.
11. State the theorem of moments (or theorem of cumulants) for linear time-invariant systems and use the theorem to find the first three moments of the impulse response of the following conceptual models:
    (a) a cascade of 2 equal linear reservoirs,
    (b) a cascade of n equal linear reservoirs,
    (c) a cascade of a pure translation and a linear reservoir,
    (d) a rectangular inflow routed through a linear reservoir,
    (e) a triangular inflow into a linear reservoir.

COMPARING MODELS

12. Compare a number of two-parameter models by plotting the dimensionless peak against the dimensionless time to peak.
13. Compare a number of two-parameter models by plotting in terms of dimensionless shape factors.
14. Fit a conceptual model to the data given in question 6. Then choose a second conceptual model and compare the ability of the two models to represent the data.
15. Repeat question 14 for the case of the data given in question 7.
16. Extend the PICOMO program given in Appendix A to include model 18 in Table 6.6: the cascade of equal linear reservoirs with uniform lateral inflow, and compare its performance with the performance of the models examined in questions 14 and 15 above.
17. Extend the programs in PICOMO to include the Linearised Channel Response (LCR) and the 2- and 3-parameter models described in Appendix D.

Acknowledgements

The proposal for this book came from Professor Pieter de Laat at the UNESCO-IHE International Institute, Delft. We would like to thank him, Marijke de Laat and Dara Dooge for their assistance and enthusiasm in bringing this project to a successful conclusion.

Professor Geoff Pegram of the University of Natal read the penultimate text and challenged both authors with his editorial comments and scientific questions. We have responded to these and thank him for his help and encouragement.

James Dooge, Dublin
J. Philip O'Kane, Cork (jpokane@ucc.ie)
December, 2002

Encomium

Since the publication occurs within the year of Jim Dooge's eightieth birthday, it is also a birthday present in his honour. The book contains the research results of very many of Jim's students, carried out over a period of forty years, at each of the three constituent colleges of the National University of Ireland at Cork, Dublin and Galway. His former students, colleagues and friends in Delft and Ireland join me in wishing him many more years of his unique engagement with the sciences of water – *ad multos annos.*

Philip O'Kane, Cork
December, 2002

References

Abbott, M.B., Bathurst, J.C., Cunge, J.A., O'Connell, P.E. and Rasmussen, J.A. (1986). An introduction to the European Hydrological System — Système Hydrologique Européen (SHE); 1. History and philosophy of a physically based distributed modelling system; 2. Structure of a physically based, distributed modelling system. *J. Hydrol.*, 87, pp. 45–59.

Abramowitz, M. and Stegun, I.A. (eds.), (1964). *Handbook of Mathematical Functions*. Applied Mathematics Series 55. U.S. National Bureau of Standards.

Ackerman, W.C., Coleman, E.A., and Ogrosky, H.O. (1955). From ocean to sky to land to ocean. In: *US Department of Agriculture Yearbook*, Washington, pp. 41–45.

Acton, F.S. (1970). *Numerical Methods that Work*. Harper and Row, New York.

Amorocho, J. (1963). Measures of the linearity of hydrologic systems. *J. Geophys. Res.*, 68, pp. 2237–2249.

Amorocho, J. and Brandstetter, A. (1971). Determination of non-linear functional response functions in rainfall-runoff processes. *Water Resour. Res.*, 7, pp. 1087–1101.

Amorocho, J. and Orlob, G.T. (1961). Non-linear analysis of hydrologic systems. Water Resources Center Contribution 40, *Univ. Calif., Berkeley*, 147 pp.

Amorocho, J. and Hart, W.E. (1964). A critique of current methods in hydrologic systems investigation. *EOS Tr. Am. Geophys. Un.*, 45, pp. 307–321.

Aravin, V.I. and Numerov, S.N. (1953). *Teoriya Dvizheny Jhidokste i Gasov i Nedieformirvemoz Porostoi Srede*. G.I.T. — T.L. Moscow. Translated into English by A. Moscona as Theory of Fluid Flow in Undeformable Porous Media. Israel Program for Scientific Translations, Jerusalem, 1965.

Aseltine, J.A. (1958). *Transform Methods in Linear System Analysis*. McGraw-Hill, New York.

Bagrov, N.A. (1953). Osrednem mnogoletnem ispraeniz c paverknosti sushi (On the average long term evaporation from the land surface). *Meteorol. Gidrol.*, 10, pp. 20–25.

Bakmeteff, B.A. (1932). *Hydraulics of Open Channels*. ASCE, New York, 329 pp.

Barnes, B.S. (1959). Consistency in unit-graphs. *Transactions of the Hydraulics Division 85(HY8)*. American Society of Civil Engineers, pp. 39–63.

Bear, J. (1972). *Dynamics of Fluids in Porous Media*. Elsevier, New York, London, Amsterdam.

Bear, J., Zaslavsky, D., and Irmay, S. (1968). Physical principles of water percolation and seepage UNESCO Arid Zone Res. XXIX, 465 pp., illus.

Becker, A. (1966). On the structure of coaxial graphical rainfall-runoff relationships. *IAHS Bull.*, 11(2), pp. 121–130.

Becker, A. and Glos, E. (1969). Grundlagen der Systemhydrologie. *Mitt. Inst. Wasserwirtsch.*

Bellman, R.E. (1961). *Adaptive Control Processes: A Guided Tour.* Princeton University Press, Princeton, New Jersey.

Benson, M.A. (1962). Factors influencing the occurrence of floods in humid regions of diverse terrain in the United States. *U.S. Geological Survey. Water-Supply Paper 530-B.* 30 pp.

Bertalanffy, L. von (1968). *General System Theory.* Braziller, New York.

Beven, K.J. and Kirkby, M.J. (1984). A physically based variable contributing area model of basin hydrology. *Hydrol. Sci. B.*, 24(1), pp. 43–69.

Bickley, W.G. and Thompson, W.A. (1964). *Matrices: Their Meaning and Manipulation.* University Press, London.

Body, D.N. (1959). *Flood Estimation: Unit Graph Procedures. Utilising a High-Speed Digital Computer.* Water Research Foundation of Australia, Bulletin 4. Sydney, Australia.

Braester, C. (1973). Moisture variation at the soil surface and the advance of the wetting front during infiltration at constant flux. *Water Resour. Res.*, 9, pp. 687–694.

Brandstetter, A. and Amorocho, J. (1970). *Generalised Analysis of Small Watershed Responses.* Water Sci. Eng., Pap. No. 1035. Dept. of Water Science and Engineering, Univ. of California, Davis, California.

Bras, R. (1990). *Hydrology, an Introduction to Hydrologic Science.* Addison-Wesley, New York.

Bras, R. and Rodriguez-Iturbe, I. (1985). *Random Functions and Hydrology.* Addison-Wesley, New York.

Brigham, E.O. (1974). *The Fast Fourier Transform.* Prentice Hall, New Jersey.

Brown, B.M. (1965). *The Mathematical Theory of Linear Systems.* Chapman and Hall, London.

Bruen, M. (1977). *Effect of Data Errors on Linear Systems Identification.* Unpublished report, Department of Civil Engineering, University College, Dublin.

Bruen, M. and Dooge, J.C.I. (1984). An efficient and robust method for estimating unit hydrograph ordinates. *J. Hydrol.*, 70, pp. 1–24.

Bruen, M. and Dooge, J.C.I. (1992a). Unit hydrograph estimation with multiple events and prior information. I: Theory and a Computer Program. *Hydrol. Sci. J.*, 37, pp. 429–443.

Bruen, M. and Dooge, J.C.I. (1992b). Unit hydrograph estimation with multiple events and prior information. II: Evaluation of the method. *Hydrol. Sci. J.*, 37, pp. 445–462.

Carnahan, B., Luther, H.A. and Wilkes, J.O. (1969). *Applied Numerical Methods.* Wiley, New York.

Carslaw, H.S. and Jaeger, J.C. (1946). *Conduction of Heat in Solids.* Oxford University Press.

Cavadias, G. (1966). *River Flow as a Stochastic Process. Symposium on Statistical Methods in Hydrology.* Government Printing Office, Ottawa, pp. 315–351.

Charny, I.A. (1951). A rigorous derivation of Dupuit's formula for unconfined seepage with a seepage surface. *Dokl. Akad. Nauk*, 79, No. 6.

Childs, E.C. (1936). The transport of water through heavy clay soils. *J. Agr. Soc.*, 26, pp. 114–141, pp. 527–545.

Childs, E.C. (1969). *An Introduction to the Physical Basis of Soil Water Phenomena.* New York, 493 pp., illus.

Chuta, P. and Dooge J. (1990). Deterministic derivation of the geomorphic unit hydrograph. *J. Hydrol.*, 147, pp. 81–97.

Ciarapica, L. and Todini, E. (1998). TOPKAPI – Un modello afflussi-deflussi applicabile della scala di versante alla scala di bacino. Proc. XXVI Convegno di Idraulica e Construzioni Idrauliche, Vol. II, pp. 49–60.

Clark, C.O. (1945). Storage and unit-hydrograph. *Tr. Am. Soc. Civil Eng.*, 110, pp. 1416–1446.

Clarke, J.I. (1966). Morphology from maps. In: G.H. Dury (ed.), *Essays in Geomorphology.* London, pp. 235–274.

Clarke, R.T. (1973). *Mathematical Models in Hydrology.* Irrigation and Drainage Paper No.19, FAO, Rome.

Clothier, B.E., Knight, J.H., and White, I. (1981). Burger's equation: application to field constant-flux infiltration. *Soil Sci.*, 132, pp. 255–261.

Collatz, L. (1966). *The Numerical Treatment of Differential Equations.* Springer-Verlag, Berlin.

Collins, W.T. (1939). Runoff distribution graph from precipitation occurring in more than one time unit. *Civil Eng.*, 9, p. 559.

Constanz, J. (1987). R.E. Moore and Yolo Light Clay. In: E.R. Landa and S. Ince (eds.), *The History of Hydrology*, American Geophysical Union, Washington, 122, pp. 99–101.

Commons, C.G. (1942). Flood hydrographs. *Civil Eng.*, New York, 12, pp. 571–573.

Corey, A.T. (1977). *Mechanics of Immiscible Fluids in Porous Media*. Water Resources Publications, Littleton, Colorado.

Corps of Engineers (1969). *Unit Hydrographs. Part I. Principles and Determinations*. U.S. Engineers Office, Baltimore.

Crawford, N.H. and Linsley, R.K. (1966). *Digital Simulation in Hydrology: Stanford Watershed Model IV.* Technical Report No. 39. Department of Civil Engineering, Stanford University, California.

Dantzig, G.B. (1963). *Linear Programming and Extensions*. Princeton, N.J. University Press.

Dawdy, D.R. and O'Donnell, T. (1965). Mathematical models of catchment behaviour. *J. Hydraul. Div.*, 91 (HY4), pp. 123–139.

Deininger, R.A. (1969). Linear programming for hydrologic analysis. *Water Resour. Res.*, 5, pp. 1105–1109.

De Laat, P.J.M. (1980). *Model for Unsaturated Flow above a Shallow Water Table*. Centre for Agricultural Publishing and Documentation, Wageningen.

De Laine, R.J. (1970). Deriving the unit hydrograph without using rainfall data. *J. Hydrol.*, 10, pp. 379–390.

De Laine, R.J. (1975). *Identifying a Time Invariant System from its Output*. PhD thesis, Monash University, Victoria.

Demidovich, B.P. and Maron, I.A. (1973). *Computational Mathematics*. Mir Publishers, Moscow.

De Neufville, R. (1990). *Applied Systems Analysis. Engineering Planning and Technology Management*. McGraw-Hill, New York.

De Wiest, J.M.R. (1966). *Geohydrology*. Wiley, New York, 366 pp., illus.

Diskin, M.H. and Boneh, A. (1972). Properties of the kernels of time-invariant, initially relaxed, second order, surface runoff systems. *J. Hydrol.*, 17, pp. 115–141.

Diskin, M.H. and Pegram, G.G.S. (1987). A study of Cell Models: 3. A Pilot Study in the Calibration of Manifold Cell Models in the Time Domain and in the Laplace Domain. *Water Resour. Res.*, 23(4), pp. 663–673.

Doebelin, E.O. (1966). *Measurement Systems: Application and Design*. McGraw-Hill, Wiley, New York.

Doetsch, G. (1961). *Guide to the Applications of Laplace Transforms*. Van Nostrand, London.

Domenico, P.A. (1972). *Concepts and Models in Groundwater Hydrology*. McGraw-Hill, New York.

Dooge, J.C.I. (1955). Discussion of The employment of unit-hydrographs to determine the flows of Irish arterial drainage channels by J.J. O'Kelly, *P. I. Civil Eng. (Ireland)*, 20, pp. 467–470.

Dooge, J.C.I. (1959). A general theory of the unit hydrograph. *J. Geophys. Res.*, 64, pp. 241–256.

Dooge, J.C.I. (1961). Discussion of A unit hydrograph study with particular reference to British catchments by J.E. Nash, *P. I. Civil Eng. (Ireland)*, 20, pp. 467–470.

Dooge, J.C.I. (1965). Analysis of linear systems by means of Laguerre functions. *SIAM J. Control*, 2, pp. 396–408.

Dooge, J.C.I. (1966). *Response of heavily damped discrete systems. Research Report*. Department of Civil Engineering. University College Cork.

Dooge, J.C.I. (1967). A new approach to non-linear problems in surface hydrology. *IAHS Publication*, 76, pp. 409–413.

Dooge, J.C.I. (1968). The hydrological cycle as a closed system. *IAHS Bull.*, 13, pp. 58–68.

Dooge, J.C.I. (1973). *Linear Theory of Hydrologic Systems*. Agriculture Research Service Technical Bulletin No. 1468, U.S. Department of Agriculture. Reprinted in 2003: EGU Reprint Series, 1, European Geosciences Union, Kattenburg, Lindau.

Dooge, J.C.I. (1977). Problems and methods of rainfall-runoff modelling. In: T. Ciriani, U. Maione, and J.R. Wallis (eds.), *Mathematics for Surface Water Hydrology*. John Wiley and Son, New York, pp. 71–108.

Dooge, J.C.I. (1979). Deterministic input–output models. In: E.H. Lloyd, T. O'Donnell and J.C. Wilkinson (eds.), *Mathematics of Hydrology and Water Resources. Proceedings of Lancaster Symposium*. Institute of Mathematics and its Application, London, pp. 1–37.

Dooge, J.C.I. (1991). Sensitivity of runoff to climate change: a Hortonian approach. Robert E. Horton Memorial Lecture. Annual Meeting of American Meteorological Society, New Orleans, January 1991. *B. Am. Meteorol. Soc.*, 73, pp. 2013–2014.

Dooge, J.C.I. and Bruen, M. (1989). Unit hydrograph stability and linear algebra. *J. Hydrol.*, 111, pp. 377–390.

Dooge, J.C.I., Bruen, M., and Parmentier, B. (1999). A simple model for estimating the sensitivity of catchment and regional runoff to long-term rainfall and long-term potential evapotranspiration. *Adv. Water Res.*, 23, pp. 153–163.

Dooge, J.C.I. and Garvey, B.M. (1978). The use of Meixner functions in the identification of heavily damped systems. *P. R. Ir. Aca., Section A*, 78, pp. 157–175.

Dooge, J.C.I. and Harley, B.M. (1967). Linear routing in uniform open channels. Int. Hydrology Symposium, Fort Collins. *Proceedings*, Vol. 1, pp. 57–63.

Dooge, J.C.I. and Napiorkowski, J.J. (1984). Effect of downstream control in flood routing. *Acta Geophys. Pol.*, 32, pp. 363–373.

Dooge, J.C.I. and Napiorkowski, J.J. (1987). Applicability of diffusion analogy in flood routing. *Acta Geophys Pol.*, 35, pp. 65–75.

Dooge, J.C.I. and Napiorkowski, J.J. (1987). The effect of the downstream boundary conditions in the linearised Saint-Venant equations. *Q. J. Mech. Appl. Math.*, 40, pp. 245–256.

Dooge, J.C.I., Napiorkowski, J.J., and Strupczewski, W.G. (1987a). The linear downstream response of a generalised uniform channel. *Acta Geophys. Pol.*, 35, No. 3. pp. 177–191.

Dooge, J.C.I., Napiorkowski, J.J., and Strupczewski, W.G. (1987b). Properties of the generalised linear downstream channel response. *Acta Geophys. Pol.*, 35, pp. 405–418.

Dooge, J.C.I. and O'Kane, J.P.J. (1977). PICOMO: A program for the identification of conceptual models. In: T. Ciriani, U. Maione, and J.R. Wallis (eds.), *Mathematical Models for Surface Water Hydrology.* John Wiley and Son, New York, pp. 277–294.

Dooge, J.C.I. and Wang, Q.J. (1993). Comment on "An investigation of the relationship between ponded and constant flux rainfall infiltration" by A. Poulavassilis, P. Kerkides, S. Elnaloglou, and I. Argyrokastritis. (*Water Resour. Res.* 27, pp. 1403–1409, 1991.) *Water Resour. Res.*, 39, pp. 1335–1337.

Edson, C.G. (1951). Parameters for relating unit hydrographs to characteristics. *Am. Geophys. Un. Trans.*, 32, pp. 591–596.

Eagleson, P.S. (1969). *Dynamic Hydrology.* McGraw-Hill, New York. Reprinted in 2003: EGU Reprint Series, 2, European Geosciences Union, Kattenburg, Lindau.

Emery, F.E. (ed.) (1969). *Systems Thinking.* Penguin Modern Management Readings. Penguin Books, Harmondsworth, England.

Erdelyi, A. (ed.) (1954). *Tables of Integral Transforms*, Vol. 1. McGraw-Hill, New York.

Faddeeva, V.N. (1959). *Computational Methods of Linear Algebra.* Translated from Russian by C.D. Benstan. Dover Publications, New York.

Fodor, G. (1965). *Laplace Transforms in Engineering.* Hungarian Academy of Sciences, Budapest.

Forrester, J.W. (1968). *Principles of Systems.* Wright Allen Press, Boston.

Fox, L. (1962). *Numerical Solution of Ordinary and Partial Differential Equations.* Pergamon, Oxford.

Franchini, M. and Pacciani, M. (1991). Comparitive analysis of several conceptual rainfall-runoff models. *J. Hydrol.*, 122, pp. 161–219.

Franklin, J.N. (1968). *Matrix Theory.* Prentice Hall, New Jersey.

Franks, R.G.E. (1967). *Mathematical Modelling in Chemical Engineering.* Wiley, New York.

Frazer, R.A., Duncan, W.J., and Collar, A.R. (1965). *Elementary Matrices and Some Applications to Dynamics and Differential Equations.* Cambridge University Press.

Freeze, R.A. (1972). Role of sub-surface flow in generating surface runoff. I. Baseflow contributions to channel flow. *Water Resour. Res.*, 8, pp. 609–623.

Fujita, H. (1952). The exact pattern of concentration dependent diffusion in a semi-infinite medium, Part II. *Text. Res. J.*, 22, pp. 823–827.

Gardner, W.R. (1958). Some steady state solution of the unsaturated moisture flow equation with applications to evaporation from a water table. *Soil Sci.*, 85, pp. 228–232

Garvey, B.M. (1972). The analysis of linear systems by means of Laguerre and Meixner functions. Thesis for M.Eng.Sc. degree. Department of Civil Engineering, University College Cork.

Gass, S.I. (1964). *Linear Programming: Methods and Application*, 2nd edition. McGraw-Hill, New York.

Glover, R.E. (1974). *Transient Groundwater Hydraulics.* Department of Civil Engineering, Colorado State University, Fort Collins, Colorado.

Glover, R.E. and Bittinger, M.W. (1959). *Source Material for a Course in Transient Groundwater Hydraulics.* Department of Civil Engineering, Fort Collins, Colorado.

Green, W.H. and Ampt, G.A. (1911). Studies of soil physics: 1 — Flow of air and water through soils. *J. Agric. Res.*, 4, pp. 1–24.

Guillemin, E.A. (1949). *The Mathematics of Circuit Analysis.* MIT Press, Cambridge, Massachusetts.

Hamming, R.W. (1962). *Numerical Methods for Scientists and Engineers.* McGraw-Hill, New York.

Hamming, R.W. (1971). *Introduction to Applied Numerical Analysis.* McGraw-Hill, New York.

Hancock (1960). *Theory of Maxima and Minima.* Dover, New York.

Harder, J.A. and Zand, S.M. (1969). *The identification of non-linear hydrologic systems.* Tech. Rep. No. HEL-8-2. Hydraul. Eng. Lab., Univ. of California, Berkeley, California.

Harr, M.E. (1962). *Groundwater and Seepage.* McGraw-Hill, New York, 315 pp., illus.

Hawken, W.H. and Ross, C.N. (1921). The calculation of flood discharges by use of a time-contour plan. *Inst. Eng. Aust. J.*, 2, pp. 85–92.

Henderson, F.M. and Wooding, R.A. (1964). Overland flow and interflow from a limited rainfall of finite duration. *J. Geophys. Res.*, 69, pp. 1531–1540.

Hino, M., Sukigara, T., and Kikkawa, T. (1971). Non-linear runoff kernels of hydrologic systems. In: *Proc. Bilateral U.S.–Jap. Seming. Hydrol. Inst.*, Water Resources Publication, Ft. Collins, Colorado, pp. 102–112.

Holtan, H.N. (1961). *A Concept for Infiltration Estimates in Watershed Engineering.* U.S. Dept. Agr. Agr. Res. Serv. ARS 41–51, 25 pp.

Horton, R.E. (1935). *Surface Runoff Phenomena. Part I. Analysis of the Hydrograph.* Horton Hydrological Laboratory Publication 101, Ann Arbor, Michigan, 73 pp.

Horton, R.E. (1938). The interpretation and application of runoff plot experiments with reference to soil erosion problems. *Soil Sci. Soc. Am. Pro.*, 3, pp. 340–349.

Horton, R.E. (1940). An approach toward a physical interpretation of infiltration capacity. *Soil Sci. Am. Pro.*, 5, pp. 399–417.

Horton, R.E. (1945). Erosional development of streams and their drainage basins: hydrophysical approach to quantitative geomorphology. *B. Geol. Soc. Am.*, 56, pp. 275–370.

Isihara, Y. and Takasao, T. (1963). Applicability of unit hydrograph method to flood prediction. *Proceedings of 10th IAHR Congress (London 1963)*, Vol. 2, pp. 81–88.

Izzard, C.F. (1944). The surface-profile of overland flow. *Am. Geophys. Un. Trans.*, 25, pp. 958–968.

Izzard, C.F. (1946). *Hydraulics of Runoff from Developed Surfaces*. Highway Res. Bd. (Washington, D.C.). Proc. 26, pp. 129–146.

Jacoby, S.L.S. (1966). A mathematical model for non-linear hydrologic systems. *J. Geophys. Res.*, 71, pp. 4811–4824.

Johnstone, D. and Cross, W.P. (1949). *Elements of Applied Hydrology*. Ronald Press, New York.

Jury, E.I. (1964). *Theory and Application of the z-Transform Method*. Wiley, New York.

Kartvelishvili, M.A. (1967). *Theory of Stochastic Processes in Hydrology and River Runoff Regulation*. Israel Program for Scientific Translations, Jerusalem.

Kaufmann, A. (1968). *Des Points et des Fleches: la Theorie des Graphes*. Dunod, Paris. Translated by H. Graham Flegg and published as *Points and Arrows: the Theory of Graphs* by Transworld Publishers Ltd. in association with Richard Sadler Ltd. London, 1972.

Kendall, M.G. and Stuart, A. (1958). *The Advanced Theory of Statistics*. Arnold, London. Vol. I, Chapter III: Moments and cumulants.

Keulegan, G.H. (1944). Spatially variable discharge over a sloping plane. *Am. Geophys. Un. Trans.*, 25, pp. 956–968.

Kirkham, D. (1966). Steady-state theories for drainage. *J. Irrig. Drain: E.–ASCE* '92(IR-I), pp. 19–39.

Kisiel, C.C. (1969). Time series analysis of hydrologic data. *Adv. Hydrosci.*, 5, pp. 1–119.

Kleitz, C. (1877). Sur la theorie du non-permanent des liquides et sur son application a la propagation des crues des rivieres. (On the theory of unsteady flow of liquids and on its application to the propagation of river floods.) Ponts et Chaussees, Annales (France, sem 2, No. 48, pp. 133–196).

Klir, G.J. (1969). *An Approach to General Systems Theory*. Van Nostrand Rheinhold, New York.

Klir, G.J. (1992). *Facets of Systems Science*. Plenum Press, New York.

Knight, J.H. and Philip, J.R. (1974). Exact solutions in nonlinear diffusion, *J. Eng. Math.*, 8, pp. 219–227.

Korn, G.A. and Korn, T.M. (1961). *Mathematical Handbook for Scientists and Engineers*. McGraw-Hill, New York.

Kostiakov, A.N. (1932). On the dynamics of the coefficients of water percolation in soils. *Sixth Commission, Int. Soc. Soil Sci.*, part A, Groningen, Moscow, pp. 15–21.

Kraijenhoff van de Leur, D.A. (1958). A study of non-steady ground-water flow with special reference to a reservoir coefficient. *Ingenieur*, 70, p. 87.

Kraijenhoff van de Leur, D.A. and others (1966). Recent trends in hydrograph synthesis. *Proceedings of Technical Meeting* 21. TNO Committee for Hydrologic Research. The Hague.

Kreider, D.L., Kuller, R.G., Ostberg, D.R., and Perkins, F.W. (1966). *An Introduction to Linear Analysis.* Addison-Wesley, New York.

Kuchment, L.S. (1967). Solution of inverse problems for linear flow problems. *Sov. Hydrol.*, 2, pp. 194–199.

Kuchment, L.S. (1972). *Matematicheskoe Modelirovanie Rechnovo Stoka.* Gidrometeorizdat, Leningrad.

Kuchment, L.S. and Borshevsky, E.N. (1971). Identification of non-linear hydrologic systems. *Meteorol. Gidrol.*, 1, pp. 42–47.

Kühnel, V. (1989). *Scale Problems in Soil Moisture Flow.* Ph.D. Thesis. National University of Ireland, Dublin.

Kühnel, V., Dooge, J.C.I., Sander, G.C. and O'Kane, J.P. (1990a). Duration of atmosphere-controlled and of soil-controlled phases of *infiltration* for constant rainfall at a soil surface. *Ann. Geophys.*, 8, pp. 11–20.

Kühnel, V., Dooge, J.C.I., Sander, G.C. and O'Kane, J.P. (1990b). Duration of atmosphere-controlled and of soil-controlled phases of *evaporation* for constant potential evaporation at a soil surface. *Ann. Geophys.*, 8, pp. 21–28.

Kundzewicz, Z.W. and Dooge, J.C.I. (1989). Attenuation and phase shift in linear flood routing. *Hydrol. Sci. J.*, 34, pp. 121–139.

Lambe, C.G. and Tranter, C.J. (1961). *Differential Equations for Engineers and Scientists.* London.

Lancaster, P. (1969). *Theory of Matrices.* Academic Press, New York.

Lanczos, C. (1957). *Applied Analysis.* Pitman, London.

Laurenson, E.M. and O'Donnell, T. (1969). Data error effects on unit hydrograph derivation. *J. Hydraul. Eng–ASCE* (HY6), pp. 1899–1917.

Lebedef, N.N. (1972). *Special Functions and their Applications* (revised edition). New York, Dover Publications.

Liggett, J.A. and Woolhiser, D.A. (1967). Difference solution of the shallow water equations. *J. Eng. Mech.–ASCE* 93 (EM-2), pp. 99 39–71.

Linsley, R.K., Kohler, M.A., and Paulhus, L.J.H. (1949). *Applied Hydrology.* McGraw-Hill, New York.

Luthin, J.N. (ed.) (1957). *Drainage of Agricultural Lands.* Agronomy 7. Madison, Wis, 620 pp., illus.

Maass, A. and others (1962). *Design of Water Resource Systems.* Harvard University Press, Cambridge, Massachusetts.

MacFarlane, A.G.J. (1964). *Engineering Systems Analysis.* Harrap, London.

Martens, H.R. and Allen, D.P. (1969). *Introduction to Systems Theory.* Merrill, Columbus, Ohio.

Meta Systems Incorporated (1975). *Systems Analysis and Water Resource Planning.* Water Information Centre, Port Washington, New York.

Minshall, N.E. (1960). Predicting storm runoff from small experimental watersheds. *J. Hydraul. Eng.–ASCE* 86 (HY-8), pp. 17–38.

Moore, R.E. (1939). Water conduction from shallow water tables. *Hilgardia.* 12, pp. 383–426.

Mulvany, T.J. (1850). On the use of self-registering rain and flood gauges. *Inst. Civil Eng. (Ireland) Trans.*, 4(2), pp. 1–8.

Muskat, M. (1937). *Flow of Homogeneous Fluids Through Porous Media.* McGraw-Hill, New York, 763 pp.

Napiorkowski, J.J. (1983). The optimisation of a third-order surface runoff model. Scientific Procedures Applied to Planning, Design and Management of Water Resources Systems. (Proc. Hamburg Symposium, August). *IAHS Publ.* 147, pp. 161–172.

Napiorkowski, J.J. (1986). Application of Volterra series to modelling of rainfall-runoff systems and flow in open channels. *Hydrolog. Sci. J.*, 31, pp. 187–203.

Napiorkowski, J.J. (1992). Linear theory of open channel flow. In: J.P. O'Kane, *Advances in Theoretical Hydrology*, Elsevier, Amsterdam, pp. 3–15.

Napiorkowski, J.J. and Dooge, J.C.I. (1988). Analytical solution of channel flow model with downstream control. *Hydrolog. Sci. J.*, 33, pp. 269–287.

Napiorkowski, J.J. and O'Kane, J.P. (1984). New non-linear conceptual model of flood waves. *Journal of Hydrology*, 69(1), pp. 43–58.

Nash, J.E. (1958). Determining runoff from rainfall. *Inst. Civil Eng. Proc.* 10, 163–184.

Nash, J.E. (1959). Systematic determination of unit hydrograph parameters. *J. Geophys. Res.*, 64. pp. 111–115.

Nash, J.E. (1960). A unit hydrographs study with particular reference to British catchments. *Inst. Civil Engin. Proc.*, 17, pp. 249–282.

Nash, J.E. (1967). The role of parametric hydrology. *Inst. Water Eng.*, 21.

Natale, L. and Todini, E. (1974). Black-box identification of a flood wave propagation linear modelling. Fifteenth Congress of IAHR. Proc, Vol. 5, pp. 165–168. Istanbul State Hydraulic Works of Turkey.

NERC (1975). Floods studies report. National Environmental Research Council, Vol. 1–5. London

Newton, D.W. and Vinyard, J.W. (1967). Computer-determined unit hydrograph from floods. *J. Hydraul. Eng.–ASCE* 93, No. HY5, pp. 219–235.

Nielsen, D.R. (1997). The vadose zone. In: N. Buras (ed.), *Reflections on Hydrology*, American Geophysical Union, Washington, pp. 205–224.

Nihoul, J.C.J. (1975). *Modelling of Marine Systems.* Elsevier Publishing Co.

O'Donnell, T. (1960). Instantaneous unit hydrograph derivation by harmonic analysis. IASH General Assembly of Helsinki. *IAHS Publ.*, 51, pp. 546–557.

O'Kelly, J.J. (1955). The employment of unit-hydrographs to determine the flow of Irish arterial drainage channels. *Inst. Civil Eng. (Ireland) Proc.*, 4, pp. 365–412.

O'Meara, W.A. (1968). *Linear Routing of Lateral Inflow in Uniform Open Channels.* M.Eng.Sc. Thesis, National University of Ireland, Cork.

Overton, D.E. (1964). *Mathematical Refinement of an Infiltration Equation for Watershed Engineering.* U.S. Dept. Agriculture ARS 41–99, 11 pp.

Papoulis, A. (1962). *The Fourier Integral and its Applications.* McGraw-Hill, New York.

Petrovskii, I.G. (1967). *Partial Differential Equations.* Iliffe, London.

Philip, J.R. (1954). An infiltration equation with physical significance. *Soil Sci. Soc. Am. Pra.*, 77, pp. 153–157.

Philip, J.R. (1957a). The theory of infiltration, Chapter 1. The infiltration equation and its solution. *Soil Sci.*, 83, pp. 345–357.

Philip, J.R. (1957b). The theory of infiltration, Chapter 4. Sorptivity and algebraic infiltration equations. *Soil Sci.*, 84, pp. 257–264.

Philip, J.R. (1968) A linearisation technique for the study of infiltration symposium on water in the unsaturated zone. *IAHS Publ.*, 82, pp. 471–478.

Philip, J.R. (1969). Theory of infiltration. In: V.T. Chow (ed.), *Advances in Hydroscience.* Academic Press, New York, pp. 219–296.

Philip, J.R. (1974). Recent progress in the solution of nonlinear diffusion equations. *Soil Sci.*, 117, pp. 257–264.

Pilgrim (1966). Radio-active tracing of storm runoff on a small catchment. *J. Hydrol.*, 4, p. 38.

Pipes, L.A. and Harvill, L.R. (1970). *Applied Mathematics for Engineers and Physicists*, 3rd edition. McGraw-Hill, New York.

Polubinorova-Kochina, P.Y. (1952). *Teoriya Dvizheniya Pochvennoi Vodi.* (Theory of groundwater movement). Translated into English by J.M.R. de Wiest. Princeton University Press.

Polya, G. (1945). *How to Solve It. A New Aspect of Mathematical Method*, 2nd edition. Princeton University Press, 253 pp.

Rainville, E.D. (1960). *Special Functions.* Macmillan, New York.

Ralston, A. and Wilf, H.S. (1960). *Mathematical Methods for Digital Computers*, Vol. I. Wiley, New York.

Raven, F.H. (1966). *Mathematics of Engineering Systems.* McGraw-Hill, New York, pp. 524.

Rektorys, K. (ed.), (1969). *Survey of Applicable Mathematics.* Iliffe, London.

Roberts, G.E. and Kaufman (1966). *Table of Laplace Transforms.* Saunders, Philadelphia.

Rodriguez-Iturbe, I. and Valdes, J.B. (1979). The geomorphologic structure of hydrologic response. *Water Resour. Res.*, 15, pp. 1409–1420.

Rogers, C., Stallybrass, M.P. and Clements, D.L. (1983). On two-phase infiltration under gravity and with boundary infiltration: an application of a Backlund transformation. *Non-linear Anal. Theory. Math. App.* 7, pp. 785–799.

Romanowicz, R.J., Dooge, J.C.I. and Kundzewicz, Z.W. (1988). Moments and cumulants of linearised St. Venant equation. *Adv. Water Res.*, 11, pp. 92–100.

Rosenbrock, H.H. and Storey, C. (1970). *Mathematics of Dynamical Systems.* Nelson, London.

Salvucci, G. (1996). Series solution for Richards equation under concentration boundary conditions and uniform initial conditions. *Water Resour. Res.*, 32, pp. 2401–2407.

Sander, G.C., Kühnel, V., Brandyk, T. Dooge, J.C. and O'Kane, J.P.J. (1986). *Analytical Solutions to the Soil Moisture Flow Equation.* Dept. of Civil Engineering, University College Dublin, EEC/NBST, Proj. CLI-038-EIR(H) Rep. No. 3.

Sander, G.C., Parlange, J.-Y., Kühnel, V., Hogarth, W.L., Lockington, D. and O'Kane, J.P.J. (1988). Exact non-linear solution for constant flux infiltration. *J. Hydrol.*, 97, pp. 341–346.

Sato, S. and Mikawa, H. (1956). A method of estimating runoff from rainfall. *Proc. Of Regional Tech. Conf. On Water Resources Development in Asia and the Far East.* UN Flood Control Series, 9: 152–155.

Saucedo, R. and Schiring, E.E. (1968). *Introduction to Continuous and Continual Control Systems.* Macmillan, New York.

Sawyer, W.W. (1972). *An Engineering Approach to Linear Algebra.* Cambridge University Press, London.

Schwartz, L. (1951). *Theorie des Distributions.* Hermann, Paris.

Seddon, A.J. (1900). River hydraulics. *Tr. Am. Soc. Civ. Eng.*, 43, pp. 179–243.

Sempere Torres, D., Rodriguez, J.Y., and Obled, Ch. (1992). Using the DPFT approach to improve flash flood forecasting models. *Nat. Hazards*, 5, pp. 17–41.

Shamseldin, A.J. and Nash, J.E. (1998) The geomorphologic unit hydrograph — Critical Review. *Hydrol. Earth Syst. Sci.*, 2(1), pp. 1–8.

Shamseldin, A.J. and Nash, J.E. (1999). Reply to Comment by G.G.S. Pegram. *Hydrol. Earth Syst. Sci.*, 3(2), pp. 311–314.

Shearer, J.L., Murphy, A.T., and Richardson, H.H. (1967). *Introduction to System Dynamics.* Addison Wesley, Reading, Massachusetts.

Sherman, L.K. (1932a). Stream flow from rainfall by the unit-graph method. *Engin. News Rec.*, 108, pp. 501–505.

Sherman, L.K. (1932b). The relation of hydrographs of runoff to size and character of drainage basins. *Am. Geophys. Un. Tr.*, 13, pp. 332–339.

Singh, V.P. (1988). *Hydrologic Systems Rainfall–Runoff Modelling.*, 2 Vols. Prentice-Hall.

Singh, V.P., Frevert, D.K. and Meyer, S.P. (2002). *Mathematical Models of Large Watershed Hydrology.* Water Resources Publications, Littleton, Colorado.

Smith, J.M. (1975). *Scientific Analysis on the Pocket Calculator.* Wiley, New York.

Sneddon, I.A. (1957). *Elements of Partial Differential Equations.* McGraw-Hill, New York.

Snyder, F.F. (1938). Synthetic unit graphs. *Am. Geophys. Un. Tr.*, 19(1), pp. 447–454.

Snyder, W.M. (1955). Hydrograph analysis by the method of least squares. *J. Hydraul. Div.–ASCE*, 81, pp. 1–25.

Snyder, W.M. (1961). *Matrix Operation in Hydrograph Computations.* TVA Research Paper No. I. Tennessee Valley Authority, Knoxville, Tennesse.

Strahler, A.N. (1964). Quantitative geomorphology of drainage basins and channel network In: Chow, V.T. (ed.), *Handbook of Applied Hydrology*, McGraw-Hill, New York, pp. 4-39–4-76.

Sugawara, M. and Maruyama, F. (1957). A method of provision of the river discharges by means of a rainfall model. Symposium Darcy. *IAHS Publ.*, 42, Vol. 3, pp. 71–76.

Svandize, G.G. (1964). *Oshovyraschata Regulirovaniya Rechnogo Stoka Metodom Monte Carlo* (Fundamental for Estimating River-flow Regulation by the Monte Carlo Method). Academy of Sciences of the Georgian Soviet Republic, Tbilisi.

Thomas, H.A. (1934). *The Hydraulics of Flood Movements in Rivers.* Carnegie Institute of Technology, Pittsburg, 70 pp.

Todini, E. (1988). *Il Modello Aflussi Deflussi de Firenze, ARNO.* General Report. Regione Toscana.

Todini, E. (1995). New Trends in modelling soil processes from hillslope to GCM scales. In: Oliver, H.R., Oliver, S.A. (eds.), *Role of Water and the Hydrological Cycle in Global Change*, Springer Verlag, New York, pp. 317–347.

Todini, E. (1996). The ARNO rainfall-runoff model. *J. Hydrol.* 175, pp. 339–382.

Todini, E. and Ciarapica, L. (2001). The TOPKAPI Model. Internal report. Department of Earth and Geo-Environmental Sciences, University of Bologna, Italy.

Todini, E. and Wallis, J.R. (1977). Using CLS for daily and longer period rainfall-runoff modelling. T. Ciriani, U. Maione and J.R. Wallis (eds.), *Mathematical Models for Surface Water Hydrology.* John Wiley and Son, New York, pp. 149–168.

Turner, J. (1982). *A Real-Time Forecasting Model for the River Liffey.* M. Eng. Sc. Thesis, University College Dublin.

Turner, J., Dooge, J.C.I., and Bree, T. (1989). Deriving the unit hydrograph by root selection. *J. Hydrol.*, 110, pp. 137–152.

Tustin, A. (1953). *Mechanism of Economic Systems.* Heinemann, London.

U.S. Corps of Engineers (1963). *Unit Hydrographs. Part I. Principles and Determinations.* U.S. Engineers Office, Baltimore.

Vemuri, V. and Vemuri, N. (1970). On the systems approach in hydrology. *Bull. IASH*, 15(2), pp. 17–38.

Wang, Q.J. (1992). Analytical and numerical modelling of unsaturated flow In: J.P. O'Kane (ed.), *Advances in Theoretical Hydrology*. Elsevier, Amsterdam, pp. 17–26.

Wang, Q.J. and Dooge, J.C.I. (1994). Limiting cases of land surface fluxes. XVI. General Assembly of European Geophysical Society, Wiesbaden, April 1991. *J. Hydrol.*, 155, pp. 429–440.

Weinberg, G.M. (1975). *An Introduction to General Systems Thinking*. Wiley, New York.

Wilde, D.J. (1964). *Optimum Seeking Method*. Prentice-Hall, Englewood Cliffs, New Jersey.

Woolhiser, David A. (1977). Unsteady free-surface flow problems. In: T. Ciriani, U. Maione, and J.R. Wallis (eds.), *Mathematical Models for Surface Water Hydrology*. John Wiley and Son, New York. pp. 195–214.

Woolhiser, D.A. and Liggett, J.A. (1967). Unsteady, one-dimensional flow over a plane — the rising hydrograph. *Water Resour. Res.*, 3, pp. 753–771.

Wiener, N. (1942). *Response of a Non-Linear Device to Noise*. Radiat. Lab. Rep. No. 129, MIT, Cambridge, Massachusetts.

Wiener, N. (1958). *Non-Linear Problems in Random Theory*. Wiley, New York.

Yevjevich, V.N. (1972). *Stochastic Processes in Hydrology*. Water Resources Publication, Fort Collins, Colorado.

Zhao Dihua and Dooge, J.C.I. (1990). A simple conceptual model of subsurface conditions. In: *Water for Life*, pp. 34–46. Proceedings of Silver Jubilee Seminar. International Centre of Hydrology. University of Padua.

Zohar, S. (1974). The solution of Toeplitz set of linear equations. *J. Ass. Comput. Mach.*, 21, pp. 272–276.

APPENDIX A

PICOMO: A Program for the Identification of Conceptual Models

A.1 INTRODUCTION

This appendix presents four different levels of description of the computer program PICOMO and its use. The program is written in an older style of FORTRAN.

Level 1 describes the program in broad terms under the headings: input data, output data and data processing. Level 2 expands these headings with sufficient detail to allow someone else to run the program. Level 3 describes the way in which the program can be used as an aid in hydrological research. Two markedly different sets of data illustrate the use of the program. Level 4 describes the program with an annotated listing.

A.2 LEVEL 1 DESCRIPTION

Input data on I/O device number 5 from the file
D:\Picomo\Picomo-b.dat

The program requires as input successive sets of rainfall-runoff pairs in any units, whether intensities or volumes. Each set of rainfall-runoff pairs must be preceded by a file header of five records identifying the set, and one record specifying (a) the number of *active* rainfall ordinates, (b) the number of runoff ordinates, and (c) the number of subdivisions of the time interval between ordinates to be used during numerical integration. Rainfall ordinates, which occur after the prescribed active number of ordinates, are assumed to contribute nothing to subsequent runoff values.

Output data on I/O device number 6 from the file
D:\Picomo\Picout-b.dat

The input data is echo-checked immediately. This is followed in separate Sections by:

(a) Normalized active-rainfall/runoff pairs and the normalizing constants; The RMS difference between normalized active-rainfall and runoff; The moments of (i) normalized active-rainfall, (ii) normalized runoff and (iii) the unit hydrograph which links (i) and (ii); The shape factors of this catchment unit hydrograph.

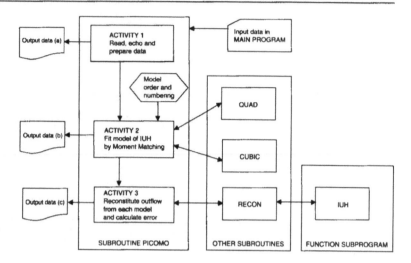

Figure A.1. The structure of PICOMO.

(b) The parameter values for the seventeen models included in the program, which are found by moment matching. The predicted shape factors not fixed by the matching and the differences between them and the shape factors of the catchment unit hydrograph.

(c) For each model, for which feasible parameter values have been found, a tableau of six columns is printed, which lists successive values in time of (i) normalized active-rainfall, (ii) the unit pulse response for the stated model, (iii) normalized runoff predicted by the stated model, (iv) normalized runoff, (v) prediction error, and (vi) time interval.

Finally the RMS error of the prediction, the volume error and the sum of the ordinates of the pulse response are printed at the bottom of each tableau. When the parameter values, which have been identified for a given model, prove to be physically unrealistic, the tableau for that model is skipped.

Data processing

The structure of the program is shown in Figure A.1. It consists of a main program calling the principal subprogram PICOMO, which is divided into three activities, which call three subroutines and one function subprogram. Communication between the sub-programs is by argument list only. There is no COMMON storage area.

Main program

The main program opens and reads the data from the input file, opens an output file and echos the text header to it and calls Subroutine PICOMO.

Subroutine PICOMO: Activity 1

In Activity 1 of *Subroutine PICOMO*, the rainfall-runoff pairs, whether they are given as intensities or volumes, are normalized to give volumes of active-rain and runoff in successive *unit* time intervals which sum to *unity*. The program is therefore 'units-free' and works in volumes. The normalizing constants are available to convert to any preferred system of units. Activity 1 then calculates and outputs the data denoted by (a) in Output data in Figure A.1. The first three moments (the second and third are about the mean) of the unit hydrograph, which is assumed to link the normalized rainfall and runoff data, are calculated using the theorem of moments and then passed to Activity 2.

Subroutine PICOMO: Activity 2

Activity 2 identifies the parameter values of each model by solving a set of equations. Models with only one parameter yield one equation in that parameter on equating the first moment of the catchment unit hydrograph derived from the data with the first moment of the model IUH. Multi-parameter models require as many moment equations as there are parameters. In general these simultaneous equations are non-linear in the parameter values.

However, the simultaneous equations arising from the two- and three-parameter models currently in PICOMO can be reduced to a single quadratic or cubic equation by elimination. The subroutines QUAD and CUBIC solve these equations explicitly. Imaginary roots, when they appear for any model, cause Activity 3 to be skipped. This test is carried out in QUAD and CUBIC. Additional tests on the parameters are carried out in Activity 2 to ensure that they are physically meaningful. For example, a negative real value for a storage delay time implies undamped exponentials in the model IUH and is therefore skipped.

When the model parameters have been fixed, the lowest order shape-factor not used in matching the output moments, is predicted. For one-parameter models, S_2 is predicted; for two-parameter models, S_3 is predicted; S_4 is not calculated. The difference is then found between the shape-factor predicted by the model and that derived from the catchment unit hydrograph. The data listed in (b) under output data in Figure A.1 is printed for each model in turn.

The internal ordering and numbering of the models in the program does not necessarily correspond to that shown on output. The order in which the models are executed, is controlled by SWITCH 1 at the start of the activity. Both the order of execution and the numbering of the models on output, can be altered by changing the initial values of the arrays ORDER (17) and REFNUM (17) in *Subroutine PICOMO*. The present initialization reproduces the order and numbering in this book (Chapter 6) and in the reference (Dooge and O'Kane, 1977) and differs from the internal program ordering.

Subroutine PICOMO: Activity 3

Activity 3 reconstitutes the normalized outflow from each model by calling and executing subroutine RECON. The activity calculates the error between the reconstituted runoff and the given runoff and outputs the data denoted by (c) in Figure A.1. The order of execution and numbering on output of the various models corresponds exactly with that in Activity 2 and is controlled by Switch 2. Any alteration to REFNUM and ORDER produces corresponding changes in Activities 1 and 2.

Subroutine RECON

Given the parameter values for a particular model, RECON calculates the ordinates of its S-curve at the specified frequency of points on the time interval between successive ordinates of rainfall and runoff (see input data). This is found by integrating numerically (trapezoidal rule) the analytical expression for the Model's IUH provided by the function sub-program IUH.

The S-curve is then "differenced" to give the ordinates of the unit pulse response, Numerical integration (trapezoidal rule) is again used to convert the sampled representation of the unit pulse response to volumes of runoff in successive intervals of time in correspondence with the rainfall and runoff data. This is called "model DUH" on output.

Finally the normalized active rainfall is convoluted with the model DUH to produce the outflow from the model.

Function sub-program IUH

IUH contains the analytical expression for the IUH of all the models considered. IUH returns to RECON the value of the IUH at a specified point in time for a given model. The model order in IUH corresponds exactly with the internal order in Activity 2. A switch at the beginning of the sub-program selects the correct IUH for the given model.

A.3 LEVEL 2 DESCRIPTION

Input description

Medium: Magnetic, ASCII text file *D:\Picomo\Picout-b.dat*.
Volume: For each data set, the number of rainfall-runoff pairs plus 2.

Record description

Record identification number	Program line	Field name(s)	Name table ref.	Format
1	6		1	'-80 cols.-' of text
2	9	NX,NY,NDELT	2,3 5	314
3	11		1	'-80 cols.-' of text
4	14	X(I), Y(I)	8,9	2F10.4 or free format

Record sequence: 1 (five times), 2 (once), 3 (once), and 4 repeated on I, NY times. A special end-of-file marker indicating no further data, is NOT used. NY controls the number of records read.

Validation of input: Echo check.

Dimension restrictions: NX < 150, NY < 150, NY * NDELT < 1000.

Instruction modification: Arrays ORDER and REFNUM of Subroutine PICOMO:

The internal order, number and title of the 17 models — 6 one-parameter, 8 two-parameter and 3 three-parameter — are

1. One linear reservoir (1 parameter)
2. Scalene triangle, peak at 1/3 the base (1 parameter)
3. Two equal linear reservoirs (ELRs) (1 parameter)
4. Single channel (1 parameter)
5. Routed rectangle (2 parameters)
6. Routed isosceles triangle (2 parameters)
7. Channel and reservoir/lag and route (2 parameters)
8. Cascade (2 ELRs) with lateral inflow (2 parameters)
9. N equal linear reservoirs, upstream inflow (2 parameters)
10. Cascade of two unequal reservoirs, upstream inflow (2 parameters)
11. Lagged cascade of N equal linear reservoirs (3 parameters)
12. Scalene triangle (2 parameter)
13. Rectangle (1 parameter)
14. Diffusion reach (1 parameter)
15. Convective diffusion reach (2 parameters)
20. Cascade of 3 unequal linear reservoirs (3 parameters)
21. Two reservoirs with unequal lateral inflow (3 parameters)

The order of fitting is specified by the index array ORDER(17) initialized in subroutine PICOMO as

```
DATA ORDER/4,13,2,1,3,14, 12,7,5,6,9,10,8,15, 11,20,21/
```

The external reference numbers, which are attached to each model on output, are contained in the index array REFNUM(17). REFNUM is initialized in subroutine PICOMO using the DATA statement

```
DATA REFNUM/1,2,4,5,6,9, 11,12,13,14,16,17,19,20, 22,23,24/
```

There is a 1 : 1 correspondence between the elements of ORDER and REFNUM. The 6 one-parameter models are fitted in the order /4,13,2,1,3,14/ and numbered externally /1,2,4,5,6,9/. The 8 two-parameter models are fitted in the order /12,7,5,6,9,10, 8,15/ and numbered externally /11,12,13,14,16,17,19,20/. The 3 three-parameter models are fitted in the order /11,20,21/ and numbered externally /22,23,24/.

This external numbering of models is the same as that used earlier in this text. REFNUM may be initialized with any set of numbers satisfying the output field I2. ORDER may be initialized with any permutation of the integers 1 to 15 (inclusive), 20 and 21. Should it be desired to execute only S ($1 < S < 17$) of the models then

(a) the first S element of ORDER must contain a subset of the numbers 1 to 15 (inclusive), 20 and 21, and
(b) the remaining elements of ORDER must *not* be members of that set of numbers.

For example, suppose we wish to execute '14. Diffusion reach' and '15. Convective diffusion reach', only, and to number them 1. and 2. on output. Then ORDER and REFNUM should be changed to read

```
DATA ORDER /14,15,0,0,0,0,0,0,0,0,0,0,0,0,0,0,0/
DATA REFNUM /1,2,0,0,0,0,0,0,0,0,0,0,0,0,0,0,0/
```

Models requiring the GAMMA function, which is not in MS-Fortran, have been removed with a zero in this way, and with comment C in the statements containing the Gamma function.

Array size modification
The dimension restrictions on NX, NY and NDELT can be relaxed by increasing the size of (a) arrays X(150), DUH(150) and PY(150) in the main program and (b) arrays X(150), S(1000), SDUH(1000), DUH(150) and PY(150) in subroutine RECON. As NY increases, the third moments will become less accurate and double precision should be considered in the calculations. If additional models are added, the number 17 must be increased accordingly wherever it occurs, e.g. in the dimensions of arrays ORDER, REFNUM, P, and in the statement $N = 17$ in subroutine PICOMO prior to the switch at the start of Activity 1. The 'big switch' at the start of Activities 1 and 2 in PICOMO, and in IUH must also be adjusted.

Output description

Medium: Magnetic, *on I/O device number 6 from file*
D:\Picomo\Picout-b.dat.
Volume: Each section (a, b, c) begins on a new page. See Figure A.1.
Section (a), 2 lines for each rainfall/runoff pair plus 50 lines. Section (b), 14 lines for each model.
Section (c), for each model executed, 11 lines plus one for each rainfall/runoff pair.

Data processing description

This has been included in the source code and is designed to facilitate the alteration and extension of the program.

Table A.1. Name table

1.	One record of 80 characters of text.
2.	NX the number of active rainfall ordinates.
3.	NY the number of runoff ordinates.
4.	DUR the duration of the time interval between successive rainfall/runoff ordinates. This always set equal to 1 in the program.
5.	NDELT the number of subdivisions of the unit time interval to be used in numerical integration.
6.	NINT = NY * NDELT the number of ordinates used in numerical integration.
7.	DT = 1/NDELT length of sub-intervals in numerical integration.
8.	X(150) the array of (normalized active) rainfall ordinates.
9.	Y(150) the array of (normalized) runoff ordinates.
10.	ORDER(17) the array of internal model numbers in the order in which they are to be executed.
11.	REFNUM(17) the reference numbers to be associated with each model on output.
12.	DUH (50) ordinates of the unit pulse response of a model as volumes in successive unit intervals of time.
13.	PY(150) normalized outflow from a model: the predicted runoff as volumes in successive unit intervals of time.
14.	S(1000) values of the S-curve for a model sampled at intervals of DT.
15.	SDUH(1000) values of the unit pulse response for a model sampled at intervals of DT.

A.4 LEVEL 3 DESCRIPTION: PROGRAM USE

Test data

The program was tested on two markedly different sets of data taken from Dooge (1973) where they are attributed to Sherman (1932) (test data set A) and to Nash (1958) (test data set B). Set A represents a complex storm and has been pre-processed to remove (a) losses from the rainfall and (b) groundwater recessions from the runoff. Set B on the other hand represents a nearly single-event storm. Furthermore, the resolution of the storm is considerably better in data set B, since its ratio of lag time to sampling interval is approximately 8, whereas it is only about 3 for A. The units for the data are shown in Table A.2. Figures A.2a and b show the echo-check and normalization of data sets A and B.

Table A.2. The test data sets.

Data set	A	B
Catchment	Big Muddy	Ashbrook
Country	USA	UK
Catchment area	753 sq. miles	248 sq.miles
Rainfall	ins	cusecs
Runoff	cusecs	cusecs
Sampling time interval	1 day	3 hrs
Lag time (approx)	3 days	24 hours
Hydrologist	Sherman	Nash

```
NUMBER OF ACTIVE RAIN ORDINATES = 16
NUMBER OF RUNOFF ORDINATES = 22
NUMBER OF SUBDIVISIONS OF TIME INTERVAL = 4

RAINFALL AND RUNOFF DATA
0.0680        132.0000
0.0000        176.0000
0.0000        229.0000
0.0000        263.0000
0.0370        312.0000
0.0000        264.0000
0.0270        167.0000
0.1020        553.0000
0.0800        666.0000
0.4400       1615.0000
0.0830       2116.0000
0.8800       4201.0000
0.6900       6188.0000
0.8200       8580.0000
0.4230      10229.0000
0.5000      11481.0000
0.0000      10924.0000
0.0400       9375.0000
0.1300       7355.0000
0.0330       5196.0000
0.0740       3500.0000
0.0250       2377.0000

NORMALISED RAINFALL AND RUNOFF DATA
0.0164          0.0015
0.0000          0.0020
0.0000          0.0027
0.0000          0.0031
0.0089          0.0036
0.0000          0.0031
0.0065          0.0019
0.0246          0.0064
0.0193          0.0078
```

Figure A.2a. Echo-check and normalization of data set A.

```
0.1060        0.0188
0.0200        0.0246
0.2120        0.0489
0.1663        0.0720
0.1976        0.0999
0.1019        0.1191
0.1205        0.1337
0.0000        0.1272
0.0000        0.1091
0.0000        0.0856
0.0000        0.0605
0.0000        0.0407
0.0000        0.0277
```

NORMALISING CONSTANTS: VX = 4.1500 VY = 85899.0000

Figure A.2a. Continued.   RMS DIFFERENCE BETWEEN X AND Y = 0.0659

NUMBER OF ACTIVE RAIN ORDINATES = 3
NUMBER OF RUNOFF ORDINATES = 38
NUMBER OF SUBDIVISIONS OF TIME INTERVAL = 5

RAINFALL AND RUNOFF DATA
```
1829.0000        0.0000
3530.0000       30.0000
8330.0000       34.0000
   0.0000      980.0000
   0.0000     1320.0000
   0.0000     1290.0000
   0.0000     1280.0000
   0.0000     1160.0000
   0.0000     1040.0000
   0.0000      910.0000
   0.0000      790.0000
   0.0000      680.0000
   0.0000      580.0000
   0.0000      480.0000
   0.0000      390.0000
   0.0000      320.0000
   0.0000      280.0000
   0.0000      240.0000
   0.0000      210.0000
   0.0000      180.0000
   0.0000      155.0000
   0.0000      135.0000
   0.0000      115.0000
   0.0000      100.0000
   0.0000       85.0000
   0.0000       70.0000
   0.0000       65.0000
   0.0000       60.0000
```

Figure A.2b. Echo-check and
normalization of data set B.

```
          0.0000          55.0000
          0.0000          50.0000
          0.0000          45.0000
          0.0000          40.0000
          0.0000          35.0000
          0.0000          30.0000
          0.0000          25.0000
          0.0000          15.0000
          0.0000           5.0000
          0.0000           0.0000
```

NORMALISED RAINFALL AND RUNOFF DATA

```
          0.1336           0.0000
          0.2579           0.0023
          0.6085           0.0026
          0.0000           0.0738
          0.0000           0.0994
          0.0000           0.0971
          0.0000           0.0964
          0.0000           0.0874
          0.0000           0.0783
          0.0000           0.0685
          0.0000           0.0595
          0.0000           0.0512
          0.0000           0.0437
          0.0000           0.0361
          0.0000           0.0294
          0.0000           0.0241
          0.0000           0.0211
          0.0000           0.0181
          0.0000           0.0158
          0.0000           0.0136
          0.0000           0.0117
          0.0000           0.0102
          0.0000           0.0087
          0.0000           0.0075
          0.0000           0.0064
          0.0000           0.0053
          0.0000           0.0049
          0.0000           0.0045
          0.0000           0.0041
          0.0000           0.0038
          0.0000           0.0034
          0.0000           0.0030
          0.0000           0.0026
          0.0000           0.0023
          0.0000           0.0019
          0.0000           0.0011
          0.0000           0.0004
          0.0000           0.0000
```

NORMALISING CONSTANTS: VX = 13689.0000 VY = 13279.0000

Figure A.2b. Continued.   RMS DIFFERENCE BETWEEN X AND Y = 0.1165

ACTIVE-RAINFALL MOMENTS
U3X = -31.0010 U2X = 6.9244 UP1X = 12.6858

RUNOFF MOMENTS
U3Y = -29.8428 U2Y = 10.8893 UP1Y = 15.9304

MOMENTS OF THE IUH
UP1H = 3.2446 U2H = 3.9649 U3H = 1.1581

SHAPE FACTORS OF THE IUH
S1 = 1.0000 S2 = 0.3766 S3 = 0.0339

Figure A.3a. Moments and shape-factors for data set A.

ACTIVE-RAINFALL MOMENTS
U3X = -0.3682 U2X = 0.5166 UP1X = 2.4749

RUNOFF MOMENTS
U3Y = 343.2381 U2Y = 37.7860 UP1Y = 10.6986

MOMENTS OF THE IUH
UP1H = 8.2237 U2H = 37.2694 U3H = 343.6063

SHAPE FACTORS OF THE IUH
S1 = 1.0000 S2 = 0.5511 S3 = 0.6178

Figure A.3b. Moments and shape-factors for data set B.

13. ROUTED RECTANGLE
BASE OF RECTANGLE = 2.8671
STORAGE DELAY TIME = 1.8111
MODEL SHAPE FACTOR S3 = 0.3478
DIFFERENCE IN S3: DATA - MODEL = -0.3139

14. ROUTED TRIANGLE
BASE OF TRIANGLE = 2.6574
STORAGE DELAY TIME = 1.9159
MODEL SHAPE FACTOR S3 = 0.4118
DIFFERENCE IN S3: DATA - MODEL = -0.3779

17. CASCADE OF TWO UNEQUAL RESERVOIRS
DELAY TIME FOR RESERVOIR 1 = 2.5150
DELAY TIME FOR RESERVOIR 2 = 0.7296
MODEL SHAPE FACTOR S3 = 0.9542
DIFFERENCE IN S3: DATA - MODEL = -0.9203

Figure A.4a. Parameters for several models of data set A.

```
19. CASCADE (2) WITH LATERAL INFLOW
DELAY TIME FOR THE RESERVOIRS = 1.4332
FRACTION OF INFLOW INTO FIRST RESERVOIR = 1.2639
MODEL SHAPE FACTOR S3 = 0.3479
DIFFERENCE IN S3: DATA - MODEL = -0.3140

20. CONVECTIVE DIFFUSION REACH
PARAMETER A = 2.0754
PARAMETER B = 0.6397
MODEL SHAPE FACTOR S3 = 0.4255
DIFFERENCE IN S3: DATA - MODEL = -0.3916
```

Figure A.4a. Continued.

```
13. ROUTED RECTANGLE
BASE OF RECTANGLE = 4.5198
STORAGE DELAY TIME = 5.9638
MODEL SHAPE FACTOR S3 = 0.7628
DIFFERENCE IN S3: DATA - MODEL = -0.1450

14. ROUTED TRIANGLE
BASE OF TRIANGLE = 4.3687
STORAGE DELAY TIME = 6.0394
MODEL SHAPE FACTOR S3 = 0.7921
DIFFERENCE IN S3: DATA - MODEL = -0.1743

17. CASCADE OF TWO UNEQUAL RESERVOIRS
DELAY TIME FOR RESERVOIR 1 = 7.6461
DELAY TIME FOR RESERVOIR 2 = 0.5776
MODEL SHAPE FACTOR S3 = 1.6082
DIFFERENCE IN S3: DATA - MODEL = -0.9904

19. CASCADE (2) WITH LATERAL INFLOW
DELAY TIME FOR THE RESERVOIRS = 4.3276
FRACTION OF INFLOW INTO FIRST RESERVOIR = 0.9003
MODEL SHAPE FACTOR S3 = 0.5826
DIFFERENCE IN S3: DATA - MODEL = 0.0352

20. CONVECTIVE DIFFUSION REACH
PARAMETER A = 2.7316
PARAMETER B = 0.3322
MODEL SHAPE FACTOR S3 = 0.9111
DIFFERENCE IN S3: DATA - MODEL = -0.2933
```

Figure A.4b. Parameters for several models of data set B.

Figure A.5a. Tableau output for model 20 (Convective-diffusion reach) of data set A.

RAINFALL	DUH-MODEL	20 PREDICTED Y	RUNOFF(Y)	ERROR	TIME
0.0163855	0.0069792	0.0001144	0.0015367	0.0014223	1
0.0000000	0.1457218	0.0023877	0.0020489	-0.0003388	2
0.0000000	0.2806440	0.0045985	0.0026659	-0.0019326	3

```
0.0000000    0.2235724    0.0036634    0.0030617   -0.0006016   4
0.0089157    0.1408795    0.0023706    0.0036322    0.0012616   5
0.0000000    0.0835226    0.0026678    0.0030734    0.0004056   6
0.0065060    0.0487890    0.0033470    0.0019441   -0.0014028   7
0.0245783    0.0284939    0.0035798    0.0064378    0.0028580   8
0.0192771    0.0167209    0.0070720    0.0077533    0.0006813   9
0.1060241    0.0098750    0.0128078    0.0188012    0.0059933  10
0.0200000    0.0058708    0.0279424    0.0246336   -0.0033088  11
0.2120482    0.0035127    0.0427768    0.0489063    0.0061295  12
0.1662651    0.0021144    0.0666471    0.0720381    0.0053910  13
0.1975904    0.0012797    0.1076292    0.0998847   -0.0077445  14
0.1019277    0.0007784    0.1370617    0.1190817   -0.0179800  15
0.1204819    0.0004756    0.1460989    0.1336570   -0.0124419  16
0.0000000    0.0002918    0.1360958    0.1271726   -0.0089232  17
0.0000000    0.0001798    0.1113842    0.1091398   -0.0022444  18
0.0000000    0.0001111    0.0735566    0.0856238    0.0120672  19
0.0000000    0.0000689    0.0443634    0.0604896    0.0161262  20
0.0000000    0.0000428    0.0261105    0.0407455    0.0146350  21
0.0000000    0.0000268    0.0153165    0.0276720    0.0123555  22
```

```
DUH TOTAL = 1.0000   PY TOTAL = 0.9776   RMS ERROR = 0.0083321
```

```
                        THE 23 TH MODEL HAS BEEN SKIPPED
                        DUE TO ILLEGAL PARAMETER VALUES.
```

Figure A.5a. Continued.

RAINFALL	DUH-MODEL	20 PREDICTED Y	RUNOFF(Y)	ERROR	TIME
0.1336109	0.0000979	0.0000131	0.0000000	-0.0000131	1
0.2578713	0.0115501	0.0015685	0.0022592	0.0006907	2
0.6085178	0.0622104	0.0113500	0.0025604	-0.0087896	3
0.0000000	0.1054652	0.0371620	0.0738007	0.0366387	4
0.0000000	0.1162651	0.0805869	0.0994051	0.0188182	5
0.0000000	0.1091116	0.1087374	0.0971459	-0.0115915	6
0.0000000	0.0957018	0.1116729	0.0963928	-0.0152801	7
0.0000000	0.0813905	0.1019498	0.0873560	-0.0145938	8
0.0000000	0.0682262	0.0883403	0.0783191	-0.0100211	9
0.0000000	0.0568222	0.0747132	0.0685293	-0.0061839	10
0.0000000	0.0472134	0.0624779	0.0594924	-0.0029854	11
0.0000000	0.0392243	0.0519931	0.0512087	-0.0007844	12
0.0000000	0.0326219	0.0432037	0.0436780	0.0004743	13
0.0000000	0.0271774	0.0359121	0.0361473	0.0002352	14
0.0000000	0.0226880	0.0298906	0.0293697	-0.0005209	15
0.0000000	0.0189818	0.0249247	0.0240982	-0.0008265	16
0.0000000	0.0159165	0.0208275	0.0210859	0.0002584	17
0.0000000	0.0133758	0.0174423	0.0180737	0.0006313	18
0.0000000	0.0112650	0.0146399	0.0158144	0.0011746	19
0.0000000	0.0095069	0.0123146	0.0135552	0.0012407	20
0.0000000	0.0080391	0.0103806	0.0116726	0.0012920	21
0.0000000	0.0068108	0.0087682	0.0101664	0.0013983	22

Figure A.5b. Tableau output for model 20 (Convective-diffusion reach) of data set B.

```
0.0000000   0.0057805   0.0074206       0.0086603   0.0012397 23
0.0000000   0.0049144   0.0062917       0.0075307   0.0012390 24
0.0000000   0.0041849   0.0053440       0.0064011   0.0010571 25
0.0000000   0.0035691   0.0045466       0.0052715   0.0007249 26
0.0000000   0.0030483   0.0038742       0.0048949   0.0010207 27
0.0000000   0.0026071   0.0033062       0.0045184   0.0012122 28
0.0000000   0.0022326   0.0028255       0.0041419   0.0013164 29
0.0000000   0.0019143   0.0024179       0.0037653   0.0013474 30
0.0000000   0.0016433   0.0020718       0.0033888   0.0013170 31
0.0000000   0.0014122   0.0017773       0.0030123   0.0012349 32
0.0000000   0.0012150   0.0015265       0.0026357   0.0011092 33
0.0000000   0.0010463   0.0013125       0.0022592   0.0009467 34
0.0000000   0.0009019   0.0011296       0.0018827   0.0007530 35
0.0000000   0.0007782   0.0009733       0.0011296   0.0001563 36
0.0000000   0.0006720   0.0008393       0.0003765  -0.0004628 37
0.0000000   0.0005808   0.0007245       0.0000000  -0.0007245 38
```

```
DUH TOTAL = 0.9962   PY TOTAL =0.9953   RMS ERROR = 0.0081580
```

Figure A.5b. Continued.

```
THE 24 TH MODEL HAS BEEN SKIPPED
DUE TO ILLEGAL PARAMETER VALUES.
```

A.5 LEVEL 4 DESCRIPTION: THE ANNOTATED PROGRAM

```
C################################################################
C################################################################
C
C  The main FORTRAN program opens and reads the data from the
C  input file, opens the output file, echos the text
C  header and calls subroutine PICOMO which fits the set
C  of conceptual models to the data.
C
C  Name table
C
C   1. One record of 80 characters of text.
C   2. NX the number of active rainfall ordinates.
C   3. NY the number of runoff ordinates.
C   4. DUR the duration of the time interval between successive
C      rainfall/runoff ordinates. This is always set equal to 1
C      in the program.
C   5. NDELT the number of subdivisions of the unit time interval
C      to be used in numerical integration.
C   6. NINT = NY * NDELT number of ordinates used in numerical
C      integration.
C   7. DT = 1/NDELT length of sub-intervals in numerical
C      integration.
C   8. X(150) the array of (normalized active) rainfall ordinates.
C   9. Y(150) the array of (normalized) runoff ordinates.
C  10. ORDER(17) array of internal model numbers in the order in
C      which they are to be executed.
C  11. REFNUM(17) the external reference numbers of each model
C      on output.
```

```
C   12. DUH (50) ordinates of the unit pulse response of a model,
C        as volumes in successive unit intervals of time.
C   13. PY(150) normalized outflow from a model: the predicted
C        runoff as volumes in successive unit intervals of time.
C   14. S(1000) values of the S-curve for a model sampled at
C        intervals of DT.
C   15. SDUH(1000) values of the unit pulse response for a model
C        sampled at intervals of DT.
C
      REAL X(150), Y(150)
      INTEGER NX,NY,NDELT
      CHARACTER TEXT(5,80)*1, TEXT1(80)*1
C
C  Open the file containing the input data
C
      OPEN (5, FILE='D:\Picomo\Picomo-a.dat')
C
C  Open a file to take the output data
C
      OPEN (6, FILE='D:\Picomo\Picout-a.dat')
C
C  Read the input file and echo the header text fields to the
C  output file
C
      READ (5,99) ((TEXT(I,J),J=1,80),I=1,5)
   99 FORMAT (80A1)
      WRITE (6,99) ((TEXT(I,J),J=1,80),I=1,5)
C
      READ (5,*) NX,NY,NDELT
  100          FORMAT (3I5)
C
      READ (5,101) (TEXT1(J),J=1,80)
  101 FORMAT (80A1)
      WRITE (6,101) (TEXT1(J),J=1,80)
C
      READ (5,*) (X(I),Y(I),I=1,NY)
  102          FORMAT (2F10.4)
C
      CLOSE (5)
C
      CALL PICOMO(X,Y,NX,NY,NDELT)
C
      CLOSE (6)
C
      END
C
C################################################################
C################################################################
C
      SUBROUTINE PICOMO(X,Y,NX,NY,NDELT)
C
C     Author: J. Philip O'Kane
C
```

```
C      Program PICOMO fits a set of conceptual models to an input-
C      output pair of arrays: x and y respectively. The array
C      x contains nx elements which represent, for example, the
C      values of effective rainfall on a catchment at successive
C      intervals of time. The array y, containing ny elements,
C      represents the corresponding ordinates of storm-runoff
C      from the catchment. Alternative uses are also possible.
C      For example x and y could represent the input and output
C      from a river-reach or the input and output from any
C      heavily damped linear system.
C      The model-fitting is accomplished by matching the first p
C      moments of the impulse response which links x and y with
C      the first p moments of the impulse response of the
C      conceptual model. PICOMO contains 17 models arranged
C      in a hierarchy of 1, 2 and 3 parameter models.
C      It attempts to fit all of them. If physically unrealistic
C      parameter values emerge from the moment matching, the
C      corresponding model is skipped, and a message is printed.
C      The internal order, number and title of the 17 models - 6
C      one parameter, 8 two parameter and 3 three parameter - are
C
C      1. One linear reservoir (1 parameter)
C      2. Scalene triangle, peak at 1/3 the base (1 parameter)
C      3. Two equal linear reservoirs (ELRs) (1 parameter)
C      4. Single channel (1 parameter)
C      5. Routed rectangle (2 parameters)
C      6. Routed isosceles triangle (2 parameters)
C      7. Channel and reservoir/lag and route (2 parameters)
C      8. Cascade (2 ELRs) with lateral inflow (2 parameters)
C      9. N equal linear reservoirs, upstream inflow (2 parameters)
C     10. Cascade of two unequal reservoirs, upstream inflow
C         (2 param)
C     11. Lagged cascade of N equal linear reservoirs (3 parameters)
C     12. Scalene triangle (2 parameter)
C     13. Rectangle (1 parameter)
C     14. Diffusion reach (1 parameter)
C     15. Convective diffusion reach (2 parameters)
C     20. Cascade of 3 unequal linear reservoirs (3 parameters)
C     21. Two reservoirs with unequal lateral inflow (3 parameters)
C
C      (Note the gap 16-19 inclusive.)
C
C         After fitting, each model is checked by comparing
C      the given and predicted values of y. The variable NDELT
C      controls the numerical integration during this step. It
C      is equal to the number of subdivisions of the time interval
C      between successive input-output ordinates. NDELT = 5 is a
C      good value to begin with.
C      Models 9 and 11 require the GAMMA function, which is not in
C      MS-Fortran. These models are suppressed in this version by
C      placing (a) zeros in the data statement ORDER below,
C      and (b) the comment C to annul the statements in Function
C      Sub-program IUH containing the Gamma function.
```

```
C
      IMPLICIT REAL(A-Z)
      REAL   DUR
      REAL   X(150),Y(150)
      REAL   VX, VY
   REAL UP1X,UP2X,UP3X,  U2X,U3X, XM1,XM2,XM3
   REAL UP1Y,UP2Y,UP3Y,  U2Y,U3Y, YM1,YM2,YM3
   REAL UP1H,U2H,U3H,  S1,S2,S3
   REAL    DUH(150),PY(150),P(5,30)
   REAL    TOTDUH,TOTPY,DIF,SSQD,RMS
      INTEGER I,J,NX,NY,NDELT,NINT,L,K,MAXO,MINO,MODEL,SKIP(30)
      INTEGER ORDER(17)
      INTEGER REFNUM(17)
      INTEGER N,RN,MOD
      DATA P/150*0.0/, DUR/1.0/, SKIP/30*0/
C
C The internal numbering of models is the sequence in ORDER,
C which is also the order in which they are fitted to data.
C
      DATA ORDER /4,13,2,1,3,14,12,7,5,6,0,10,8,15, 0,20,21/
C                                         9               11
C To reinstate models 9 and 11 replace the zeros with the
C numbers 9 and 11 above, and.
C remove the two comments C in REAL FUNCTION IUH(T,MODEL,P)
C following statements 9 and 11 in IUH
C to reactivate the expressions for their IUHs
C     IUH = (H**(LRNN-1.0))*EXP(-H)/(LRNK*GAMMA(LRNN))
C     IUH = (H**(LCNN-1.0))*EXP(-H)/(LCNK*GAMMA(LCNN))
C
C The external numbering on output is the sequence
C in REFNUM corresponding to that in ORDER
C
      DATA REFNUM /1,2,4,5,6,9,11,12,13,14,16,17,19,20,22,23,24/
C
C Use the following replacements to run models 14 and 15 only
C and to assign the reference numbers 1 and 2 to them on output.
C
C DATA ORDER /14,15,0,0,0,0,0,0,0,0,0,0,0,0,0,0,0/
C DATA REFNUM /1,2,0,0,0,0,0,0,0,0,0,0,0,0,0,0,0/
C
C*****************************************************************
C ACTIVITY 1 ECHO AND PREPARE DATA.
C*****************************************************************
C
     WRITE (6,101) NX,NY,NDELT
 101 FORMAT (1H ,//,' NUMBER OF ACTIVE RAIN ORDINATES =',I4//'
    NUMBER 1 OF RUNOFF ORDINATES =',I4//' NUMBER OF SUBDIVISIONS
    OF TIME INTER 2VAL =',I4)
     WRITE (6,103) (X(I),Y(I),I = 1,NY)
 102 FORMAT (2F10.4)
 103 FORMAT (1H ,//' RAINFALL AND RUNOFF DATA',//( F10.4,2X,F10.4))
C
C  SET RAINFALL TO ZERO BEYOND NX
```

```
      C
            J = NX + 1
            DO 20 I = J,NY
         20 X(I) = 0.0
      C
      C                         NORMALISE DATA
      C
            VX = 0.0
            VY = 0.0
            DO 10 I = 1,NY
            VX = VX + X(I)
         10 VY = VY + Y(I)
            DO 11 I = 1,NX
         11 X(I) = X(I)/VX
            DO 12 I = 1,NY
         12 Y(I) = Y(I)/VY
      C
      C  The DO loop terminating on statement 10 accumulates the
      C  individual rain ordinates in VX and the runoff
      C  ordinates in VY. The loops on 11 and 12, divide all
      C  the X and Y values by these  totals, normalizing the
      C  data pairs to fractions of unity. The moment theorem
      C  requires this.
      C
            WRITE (6,104)
        104 FORMAT(1H ,/,7X,'NORMALISED')
            WRITE (6,111) (X(I),Y(I),I=1,NY)
        111 FORMAT (1H ,/' RAINFALL AND RUNOFF DATA',//(F10.4,2X,F10.4))
            WRITE(6,120) VX,VY
        120 FORMAT(1H ,/,'  NORMALISING CONSTANTS: VX =',F14.4,'  VY =',
           1 F14.4)
            SSQD = 0.0
            DO 21 I = 1,NY
            DIF = Y(I) - X(I)
         21 SSQD = SSQD + DIF*DIF
            RMS = (SSQD/NY)**0.5
      C
      C  The sum of the squares of the differences SSQD is found between
      C  all the Ys and Xs and the RMS difference is calculated.
      C
            WRITE(6,121) RMS
        121 FORMAT(1H ,/,'   RMS DIFFERENCE BETWEEN X AND Y =',F10.4)
      C
      C                     CALCULATE MOMENTS
      C
            UP1X = 0.0
            UP2X = 0.0
            UP3X = 0.0
            DO 13 I = 1,NX
            XM1 = I*DUR
            UP1X = UP1X + XM1*X(I)
            XM2 = XM1*XM1
            UP2X = UP2X + XM2*X(I)
```

```
      XM3 = XM2*XM1
   13 UP3X = UP3X + XM3*X(I)
C
C  We now accumulate the first three moments of the rainfall data
C  about the time origin in the loop to 13 on I.
C  The first moment is UP1X; U stands for Greek mu for moment;
C  P is for prime for moments about t = 0; 1 is for first
C  moment of X.
C  The other moments have related mnemonics UP2X, and UP3X.
C  The moment arm is I*DUR where DUR = 1 is the time interval of
C  duration 1 between ordinates.
C
C
      U2X = UP2X - UP1X**2
      U3X = UP3X - 3.0*UP1X*UP2X + 2.0*(UP1X**3)
C
C  The two statements immediately above convert moments
C  about t = 0 to moments about t = UP1X the mean.
C  Hence these two moments U2X and U3X have no P for prime.
C
      WRITE (6,105) U3X,U2X,UP1X
  105 FORMAT (1H ,//,'ACTIVE-RAINFALL MOMENTS',//,'
      U3X =',F10.4,4X,  1'U2X =',F10.4,4X,'UP1X =',F10.4)
C
C  The runoff moments UP1Y etc are now found in exactly the
C  same way.
C
      UP1Y = 0.0
      UP2Y = 0.0
      UP3Y = 0.0
      DO 23 I = 1,NY
      YM1 = I*DUR
      UP1Y = UP1Y + YM1*Y(I)
      YM2 = YM1*YM1
      UP2Y = UP2Y + YM2*Y(I)
      YM3 = YM2*YM1
   23 UP3Y = UP3Y + YM3*Y(I)
      U3Y = UP3Y - 3.0*UP1Y*UP2Y + 2.0*(UP1Y**3)
      U2Y = UP2Y - UP1Y**2
      WRITE (6,110) U3Y,U2Y,UP1Y
  110 FORMAT (1H ,//,'RUNOFF MOMENTS',//,' U3Y =',F10.4,4X,
      1'U2Y =',F10.4,4X,'UP1Y =',F10.4)
C
C  MOMENTS OF IUH
C
C  The moments of the unknown IUH, which is presumed to link X
C  and Y, are found by subtraction in accordance with the
C  theorem of moments.
C  UP1H is the first (1) moment (U) about the origin (P) of the
C  IUH (H) or the lag time for the catchment.
C  The second and third moments U2H and U3H are about t = UP1H.
C
      UP1H = UP1Y - UP1X
```

```
      U2H = U2Y - U2X
      U3H = U3Y - U3X
C
C                    DUR CORRECTIONS
C
C     We can also apply Sheppard type DUR corrections assuming
C  we have a truely pulsed input and a sampled output.
C  These can be activated by removing the Cs in the first
C  column below.
C  Corresponding changes are required.at statement 216 in
C  subroutine RECON.
C
C    UP1H=UP1Y-UP1X-DUR/2.0
C    U2H=U2Y-U2X-(DUR*DUR)/12.0
C    U3H=U3Y-U3X
C
C                    SHAPE FACTORS OF IUH
C
      S1 = 1.0
      S2 = U2H/(UP1H**2)
      S3 = U3H/(UP1H**3)
C
C  The dimensionless shape factors S2 and S3 are also found.
C
      WRITE(6,106) UP1H,U2H,U3H,S1,S2,S3
  106 FORMAT(1H ,//,'MOMENTS OF THE IUH',//,' UP1H =',F12.4,2X,
     'U2H =', 1F12.4,2X,'U3H =',F12.4,//'SHAPE FACTORS OF THE
     IUH',//' S1 =', 2F12.4,2X,' S2 =',F12.4,2X,' S3 =',F12.4,//)
C
C  All quantities are written to the output file.
C  This completes Activity 1.
C
C*****************************************************************
C ACTIVITY 2 FIT MODELS OF IUH BY MOMENT MATCHING
C*****************************************************************
C  In Activity 2 we fit 17 (the value of N) different conceptual
C  models by matching the moments of their analytical IUHs with
C  the numerical IUH moments from the (X-Y) data set.
C
C  All 17 models are handled through ORDER (I) in the same way.
C  When the moment equations do not have an explicit solution,
C  they have been reduced to one quadratic or one cubic equation
C  which is solved through a call to subroutine QUAD or CUBIC.
C  These subroutines contain the standard formulae for these
C  equations.
C
C  The DO loop which drives activity 2 terminates on the
C  799 CONTINUE statement.
C  Activity 2 contains a switch just inside the DO loop to
C  statement 799 at the end of this activity. As each model is
C  executed a jump is made, GO TO 798, to the 798 CONTINUE
C  statement which immediately precedes 799 CONTINUE.
C
```

```
C  In all models, physically unrealistic parameter values can
C  occur.
C  The SKIP variable and explicit tests in QUAD, CUBIC and the
C  main program trap occurrences. The array P collects the
C  parameter values for each model.
C
      N = 17
C.................................................................
C    SWITCH 1 BEGIN
C.................................................................
      DO 799 I = 1,N
      MODEL = ORDER(I)
      RN = REFNUM(I)
      IF(MODEL.EQ.0) GO TO 798
      GO TO (701,702,703,704,705,706,707,708,709,710), MODEL
      MOD = MODEL - 10
      GO TO (711,712,713,714,715), MOD
      MOD = MOD - 9
      GO TO (720,721),MOD
      N = N - 1
      GO TO 798
C
C  The index I of the DO loop, picks from ORDER(I), the order in
C  which the models are to be fitted. For example, when I = 11, it
C  picks out ORDER(11), which is 9, and assigns 9 to MODEL. Its
C  reference number on output is REFNUM(11) which is
C  16 (external order).
C  The multiway branching GO TO then transfers control to
C  statement 709, which starts the set of statements for fitting
C  model 9 (internal order) i.e. Model 16: the Nash cascade.
C  The other models are treated in the same way. The WRITE
C  statement in each case provides the external model number
C  RN = REFNUM(I), model name and the model parameters.
C.................................................................
C    SWITCH 1 END
C.................................................................
C
C         1. ONE LINEAR RESERVOIR
C
C  LR stands for one linear reservoir. The appendage K refers
C  to the single parameter LRK for this model. The first
C  statement calculates LRK in order to match exactly the first
C  moment of the model, with the first moment calculated from the
C  (X-Y) data, namely, LRK = UP1H.
C  Since the second order moment is not matched,
C  the dimensionless second moment for the fitted model is
C  calculated as LRS2 and compared with S2 from the (X-Y)
C  data. The difference between the two, DLRS2,is also found
C  and the results written out. The same pattern
C  is followed in all of the other models.
C
  701 LRK = UP1H
      LRS2 = 1.0
```

```
            DLRS2 = S2 - LRS2
            WRITE (6,107) RN,LRK,LRS2,DLRS2
            P(1,1) = LRK
        107 FORMAT(////,I3,'. ONE LINEAR RESERVOIR',
           2//,' STORAGE DELAY TIME =',F10.4,
           3//,' MODEL SHAPE FACTOR S2 =',F10.4,
           4//,' DIFFERENCE IN S2: DATA - MODEL =',F10.4)
            GO TO 798
    C
    C           2. SCALENE TRIANGLE (Peak at 1/3 point)
    C
        702 BTR = (9.0*UP1H)/4.0
            TRS2 = 7.0/32.0
            DTRS2 = S2 - TRS2
            WRITE (6,108) RN,BTR,TRS2,DTRS2
            P(1,2) = BTR
        108 FORMAT(////,I3,'. TRIANGLE (peak at 1/3 point)',
           1//,' BASE OF TRIANGLE =',F10.4,
           2//,' MODEL SHAPE FACTOR S2 =',F10.4,
           3//,' DIFFERENCE IN S2: DATA - MODEL =',F10.4)
            GO TO 798
    C
    C           3. TWO EQUAL LINEAR RESERVOIRS
    C
        703 LR2K = UP1H/2.0
            LR2S2 = 0.5
            DLR2S2 = S2 - LR2S2
            WRITE (6,109) RN,LR2K,LR2S2,DLR2S2
            P(1,3) = LR2K
        109 FORMAT(////,I3,'. TWO EQUAL LINEAR RESERVOIRS',
           2//,' STORAGE DELAY TIME =',F10.4,
           3//,' MODEL SHAPE FACTOR S2 =',F10.4,
           4//,' DIFFERENCE IN S2: DATA - MODEL =',F10.4)
            GO TO 798
    C
    C           4. SINGLE CHANNEL
    C
        704 SCT = UP1H
            SCS2 = 0.0
            DSCS2 = S2 - SCS2
            WRITE(6,117) RN,SCT,SCS2,DSCS2
            DT = DUR/NDELT
            P(1,4) = SCT
            P(2,4) = DT
        117 FORMAT(////,I3,'. SINGLE CHANNEL',
           1//,' DELAY TIME FOR THE CHANNEL =',F10.4,
           2//,' MODEL SHAPE FACTOR S2 =',F10.4,
           3//,' DIFFERENCE IN S2: DATA - MODEL =',F10.4)
            GO TO 798
    C
    C           5. ROUTED RECTANGLE
    C
        705 A = 1.0/3.0
```

```
          B = -UP1H
          C = UP1H*UP1H - U2H
          CALL QUAD (A,B,C,RRT,SKIP(5))
          RRK = UP1H - RRT/2.0
          IF (RRK.LE.0.0) SKIP(5) =1
          RRS3 = (2.0*RRK**3)/((RRT/2.0 + RRK)**3)
          DRRS3 = S3-RRS3
          WRITE(6,113) RN,RRT,RRK,RRS3,DRRS3
          P(1,5) = RRT
          P(2,5) = RRK
      113 FORMAT (////,I3,'. ROUTED RECTANGLE',
         1//,' BASE OF RECTANGLE =',F10.4,
         2//,' STORAGE DELAY TIME =',F10.4,
         3//,' MODEL SHAPE FACTOR S3 =',F10.4,
         4//,' DIFFERENCE IN S3: DATA - MODEL =',F10.4)
          GO TO 798
C
C          6. ROUTED ISOSCELES TRIANGLE
C
      706 A = 7.0/24.0
          B = -UP1H
          C = UP1H*UP1H - U2H
          CALL QUAD (A,B,C,RTT,SKIP(6))
          RTK = UP1H - RTT/2.0
          IF (RTK.LT.0.0) SKIP(6) = 1
          RTS3 = (2.0*RTK**3)/((RTT/2.0 + RTK)**3)
          DRTS3 = S3 - RTS3
          WRITE(6,114) RN,RTT,RTK,RTS3,DRTS3
          P(1,6) = RTT
          P(2,6) = RTK
      114 FORMAT (////,I3,'. ROUTED ISOSCELES TRIANGLE',
         1//,' BASE OF TRIANGLE =',F10.4,
         2//,' STORAGE DELAY TIME =',F10.4,
         3//,' MODEL SHAPE FACTOR S3 =',F10.4,
         4//,' DIFFERENCE IN S3: DATA - MODEL =',F10.4)
          GO TO 798
C
C          7. CHANNEL AND RESERVOIR / LAG AND ROUTE
C
      707 A = 1.0
          B = -2.0*UP1H
          C = UP1H*UP1H - U2H
          CALL QUAD(A,B,C,CRT,SKIP(7))
          CRK = UP1H - CRT
          IF (CRK.LT.0.0) SKIP(7) = 1
          CRS3 = (2.0*CRK**3)/((CRT+CRK)**3)
          DCRS3 = S3 - CRS3
          WRITE(6,115) RN,CRT,CRK,CRS3,DCRS3
          P(1,7) = CRT
          P(2,7) = CRK
      115 FORMAT(////,I3,'. CHANNEL AND SINGLE LINEAR RESERVOIR',
         1//,' DELAY TIME FOR THE CHANNEL =',F10.4,
         2//,' STORAGE DELAY TIME =',F10.4,
```

```
                3//,' MODEL SHAPE FACTOR S3 =',F10.4,
                4//,' DIFFERENCE IN S3: DATA - MODEL =',F10.4)
                GO TO 798
      C
      C            8. CASCADE ( 2 E.L.R.S ) WITH LATERAL INFLOW
      C
        708 A = 2.0
            B = -4.0*UP1H
            C  = U2H + UP1H*UP1H
            CALL QUAD (A,B,C,CLIK,SKIP(8))
            BETA = (UP1H - CLIK)/CLIK
      C     IF(BETA .LE.0.0.OR.BETA .GE.1.0) GO TO 18
            CLIS3 = 2.0*(2.0-(1.0-BETA )**3)/(1.0+BETA )**3
            GO TO 19
         18 SKIP(8) = 1
            CLIS3 = 0.000001
         19 DCLIS3 = S3 - CLIS3
            WRITE(6,116) RN,CLIK,BETA ,CLIS3,DCLIS3
            P(1,8) = CLIK
            P(2,8) = BETA
        116 FORMAT(////,I3,'. CASCADE (2) WITH LATERAL INFLOW',
            1//,' DELAY TIME FOR THE RESERVOIRS =',F10.4,
            2//,' FRACTION OF INFLOW INTO FIRST RESERVOIR =',F10.4,
            3//,' MODEL SHAPE FACTOR S3 =',F10.4,
            4//,' DIFFERENCE IN S3: DATA - MODEL =',F10.4)
            GO TO 798
      C
      C
      C            9. N EQUAL LINEAR RESERVOIRS (THE NASH CASCADE)
      C
      C     In model 9 (The Nash cascade)
      C  LRN stands for n linear reservoirs. The appendages N and K
      C  refer to the two parameters LRNN and LRNK for this model.
      C  The first two statements calculate their values to match
      C  exactly the first two model moments with the moments
      C  calculated from the (X-Y) data, namely, UP1H and U2H.
      C  Since the third and higher order moments are not matched,
      C  the dimensionless third moment for the fitted model is
      C  calculated as LRNS3 and compared with S3 from the (X-Y) data.
      C  The difference between the two is also found and the results
      C  written to the output file.
      C
        709 LRNK = U2H/UP1H
            LRNN = (UP1H*UP1H)/U2H
            LRNS3 = 2.0/(LRNN*LRNN)
            DLRNS3 = S3 - LRNS3
            IF (LRNK.LE.0.0.OR.LRNN.LE.0.0) SKIP(9) = 1
            WRITE(6,112) RN,LRNN,LRNK,LRNS3,DLRNS3
            P(1,9) = LRNK
            P(2,9) = LRNN
        112 FORMAT(////,I3,'. N EQUAL LINEAR RESERVOIRS',
            1//,' NUMBER OF RESERVOIRS IN CASCADE =',F10.4,
            2//,' STORAGE DELAY TIME FOR EACH RES. =',F10.4,
```

```
         3//,' MODEL SHAPE FACTOR S3 =',F10.4,
         4//,' DIFFERENCE IN S3: DATA - MODEL =',F10.4)
         GO TO 798
C
C          10. CASCADE OF TWO UNEQUAL RESERVOIRS
C
  710 A = 2.0
         B = -2.0*UP1H - U2H
         C = UP1H*UP1H - U2H
         CALL QUAD(A,B,C,C2RK2,SKIP(10))
         C2RK1 = UP1H - C2RK2
         IF(C2RK1.LT.0.0) SKIP(10) = 1
         IF(C2RK1.EQ.C2RK2) SKIP(10) = 1
         C2RS3 = (2.0*C2RK1**3 + 2.0*C2RK2**3)/((C2RK1+C2RK2)**3)
         DC2RS3 = S3 - C2RS3
         P(1,10) = C2RK1
         P(2,10) = C2RK2
         WRITE(6,118) RN,C2RK1,C2RK2,C2RS3,DC2RS3
  118 FORMAT(////,I3,'. CASCADE OF TWO UNEQUAL RESERVOIRS',
         1//,' DELAY TIME FOR RESERVOIR 1 =',F10.4,
         2//,' DELAY TIME FOR RESERVOIR 2 =',F10.4,
         3//,' MODEL SHAPE FACTOR S3 =',F10.4,
         4//,' DIFFERENCE IN S3: DATA - MODEL =',F10.4)
         GO TO 798
C
C          11. LAGGED CASCADE OF N EQUAL LINEAR RESERVOIRS
C
  711 LCNK = U3H/(2.0*U2H)
         IF(LCNK.LT.0.0001) SKIP(11) = 1
         LCNN = (4.0*U2H**3)/(U3H**2)
         IF(LCNN.LT.0.0001) SKIP(11) = 1
         LCNT = UP1H - (2.0*U2H**2)/U3H
         IF(LCNT.LT.0.0001) SKIP(11) = 1
         P(1,11) = LCNK
         P(2,11) = LCNN
         P(3,11) = LCNT
         WRITE(6,119) RN,LCNK,LCNN,LCNT
  119 FORMAT(////,I3,'. LAGGED CASCADE OF N EQUAL RESERVOIRS',
         1//,' STORAGE DELAY TIME FOR EACH RES. =',F10.4,
         2//,' NUMBER OF RESERVOIRS IN CASCADE =',F10.4,
         3//,' DELAY TIME BEFORE THE CASCADE','=',F10.4)
         GO TO 798
C
C          12. SCALENE TRIANGLE
C
  712 A = 2.0*S2-1.0
         B = 4.0*S2+1.0
         C = 2.0*S2-1.0
         CALL QUAD(A,B,C,AB,SKIP(12))
         BTR = 3.0*UP1H/(1.0+AB)
         SUB = 9.0*AB/(1.0+AB)**2
         ITS3 = (2.0 - SUB)/10.0
         DITS3 = S3 - ITS3
```

```
          WRITE(6,122) RN,AB,BTR,ITS3,DITS3
          P(1,12) = AB
          P(2,12) = BTR
     122 FORMAT(/////,I3,'. SCALENE TRIANGLE',
        1//,' PEAK TIME AS A FRACTION OF THE BASE =',F10.4,
        2//,' BASE OF TRIANGLE =',F10.4,
        3//,' MODEL SHAPE FACTOR S3 =',F10.4,
        4//,' DIFFERENCE IN S3: DATA - MODEL =',F10.4)
          GO TO 798
C
C          13. RECTANGLE
C
     713 BR = 2.0*UP1H
          RS2 = 0.3333333
          DRS2 = S2 - RS2
          WRITE(6,123) RN,BR,RS2,DRS2
          P(1,13) = BR
     123 FORMAT(/////,I3,'. RECTANGLE',
        1//,' BASE OF RECTANGLE',F10.4,
        3//,' MODEL SHAPE FACTOR S2 =',F10.4,
        4//,' DIFFERENCE IN S2: DATA - MODEL =',F10.4)
          GO TO 798
C
C          14. DIFFUSION REACH
C
     714 A = (UP1H**3/(2.0*U2H))**0.5
          WRITE(6,124) RN,A
          P(1,14) = A
     124 FORMAT(/////,I3,'. DIFFUSION REACH',
        1//,' PARAMETER A =',F10.4)
          GO TO 798
C
C
C          15. CONVECTIVE DIFFUSION REACH
C
     715 A = (UP1H**3/(2.0*U2H))**0.5
          B = (UP1H/(2.0*U2H))**0.5
          CDS3 = 0.75/(A*A*B*B)
          DCDS3 = S3 - CDS3
          WRITE(6,125) RN,A,B,CDS3,DCDS3
          P(1,15) = A
          P(2,15) = B
     125 FORMAT(/////,I3,'. CONVECTIVE DIFFUSION REACH',
        1//,' PARAMETER A =',F10.4,
        2//,' PARAMETER B =',F10.4,
        3//,' MODEL SHAPE FACTOR S3 =',F10.4,
        4//,' DIFFERENCE IN S3: DATA - MODEL =',F10.4)
          GO TO 798
C
C          20. CASCADE OF 3 RESERVOIRS
C
     720 D = -(UP1H**3 - 3.0*UP1H*U2H + U3H)/6.0
          C = (UP1H**2 - U2H)/2.0
```

```
      B = -UP1H
      A = 1.0
      CALL CUBIC(A,B,C,D,K1,K2,K3,SKIP(20))
      IF(K1.LT.0.001.OR.K2.LT.0.001.OR.K3.LT.0.001) SKIP(20)=1
      WRITE(6,126) RN,K1,K2,K3
      P(1,20) = K1
      P(2,20) = K2
      P(3,20) = K3
  126 FORMAT(////,I3,'. CASCADE OF 3 RESERVOIRS',
     1//,' DELAY TIME FOR RESERVOIR 1 =',F10.4,
     2//,' DELAY TIME FOR RESERVOIR 2 =',F10.4,
     3//,' DELAY TIME FOR RESERVOIR 3 =',F10.4)
      GO TO 798
C
C  21. TWO RESERVOIRS WITH LATERAL INFLOW
C
  721 UP2H = U2H + UP1H*UP1H
      UP3H = U3H + 3.0*UP1H*U2H + UP1H**3
      A = UP1H**2 - UP2H/2.0
      B = UP3H/6.0 - UP1H*UP2H/2.0
      C = (UP2H/2.0)**2 - UP1H*UP3H/6.0
      CALL QUAD(A,B,C,K2,SKIP(21))
      K1 = C
      IF(K1.LT.0.001.OR.K2.LT.0.001) SKIP(21)=1
      BETA = (UP1H - K2)/K1
      P(1,21) = BETA
      P(2,21) = K1
      P(3,21) = K2
      WRITE(6,127) RN,K1,K2,BETA
  127 FORMAT(////,I3,'. TWO RESERVOIRS WITH LATERAL INFLOW',
     1//,' DELAY TIME FOR RESERVOIR 1 =',F10.4,
     2//,' DELAY TIME FOR RESERVOIR 2 =',F10.4,
     3//,' FRACTION OF INFLOW INTO FIRST RES. =',F10.4)
  798 CONTINUE
  799 CONTINUE
C
C    This completes Activity 2
C
C**********************************************************************
C ACTIVITY 3 RECONSTITUTE OUTFLOW USING EACH MODEL
C                    AND OUTPUT RESULTS
C**********************************************************************
C
C  Activity 3 reconstitutes the outflow using the models which
C  were fitted in activity 2. A loop on J runs from 1 to (N=)17
C  and terminates at 699 at the end of the activity. ORDER (J)
C  again controls the order in which this is done. For example,
C  if J = 11, MODEL = 9 and RN = 16.
C.....................................................................
C    SWITCH 2 BEGIN
C.....................................................................
      DO 699 J = 1,N
      MODEL = ORDER(J)
```

```
          RN = REFNUM(J)
C..............................................................
C    SWITCH 2 END
C..............................................................
C If physically realistic values of the parameters have been
C found, control is passed to SUBROUTINE RECON, otherwise
C MODEL is skipped.
C
     IF(MODEL.EQ.0) GO TO 698
       IF(SKIP(MODEL).EQ.1) GO TO 698
C
C The argument list of
C
C SUBROUTINE RECON (NX, NY, DUR, NDELT, X, MODEL, P, DUH, PY)
C
C transfers essential data to and from subprograms PICOMO and
C RECON.
C The inputs to RECON are the first seven arguments of the list.
C The last two arguments contain the output from RECON.
C DUH contains the ordinates of the unit pulse repsonse for
C MODEL when the corresponding parameters are P. The array PY
C contains the predicted output from the model when the
C input is X.
C
     CALL RECON(NX,NY,DUR,NDELT,X,MODEL,P,DUH,PY)
C
C The results header for each model is written to the
C output file.
C
     WRITE(6,1110) RN
1110 FORMAT(////,1H ,' RAINFALL DUH-MODEL ',I2,' PREDICTED Y
     RUNOFF(1Y) ERROR TIME',/)
C
C DIF is the difference between the given output and the
C predicted output at successive points in time.
C These are squared and accumulated in SSQD.
C Likewise, the ordinates of DUH are accumulated in TOTDUH.
C RMS is the root mean squared error. TOTDUH should
C be one. If TOTDUH is less than one, there is a
C tail to the DUH, which has not been included in PY beyond
C PY(NY), and the total predicated output is less than the total
C unit input.
C The value of TOTPY is the fraction of the total unit input
C reproduced in the predicted output PY. This distorts
C the comparison with other models exhibiting TOTDUH=1
C which are based on RMS or other error norms.
C If TOTDUH is greater than one, it should be investigated.
C
          TOTDUH = 0.0
          TOTPY=0.0
          SSQD = 0.0
     DO 199 I = 1,NY
          DIF = Y(I) - PY(I)
```

```
C
C  The results for each model are written to the output file.
C
      WRITE(6,1114) X(I),DUH(I),PY(I),Y(I),DIF,I
      1114 FORMAT(1H ,5(F10.7,2X),I3)
      TOTPY = TOTPY + PY(I)
      TOTDUH = TOTDUH + DUH(I)
      199 SSQD = SSQD + DIF*DIF
C
      RMS = (SSQD/NY)**0.5
C
C  TOTDUH, TOTDIF and RMS errors for each model are written
C  to the output file.
C
           WRITE(6,1115) TOTDUH, TOTPY, RMS
 1115 FORMAT(1H ,/,'DUH TOTAL =',F6.4,' PY TOTAL =',F6.4,
      1'    RMS ERROR =',F10.7,/////)
C
      GO TO 699
C
C
  698 SKIP(MODEL) = 0
      WRITE(6,130) RN
  130 FORMAT(/////,10X,'THE',I3,' TH MODEL HAS BEEN SKIPPED ',
      1/,10X,'DUE TO ILLEGAL PARAMETER VALUES.',/////)
C
  699 CONTINUE
  997 RETURN
      END
C
C##################################################################
C
      SUBROUTINE QUAD (A,B,C,P,SKIP)
      INTEGER SKIP
C
C  P is the physically meaningful root of A.X**2 + B.X + C = 0
C  If none is found for the given data (A,B,C), SKIP is set to 1
C  and P is given a very small value.
C
      SKIP = 0
      DELSQ = B*B - 4.0*A*C
      IF(DELSQ.LT.0.0) GO TO 5
C
      DEL = DELSQ**0.5
      R1 = (-B+DEL)/(2.0*A)
      R2 = (-B-DEL)/(2.0*A)
C
      IF (R1.LT.0.0) GO TO 2
      IF (R2.LT.0.0) GO TO 3
C
      P = AMIN1(R1,R2)
      C = AMAX1(R1,R2)
      RETURN
```

```
C
  2 IF (R2.LT.0.0) GO TO 4
    P = R2
    RETURN
C
  3 P = R1
    RETURN
C
  4 SKIP = 1
    P = 0.0002
    RETURN
C
  5 SKIP = 1
    P = 0.000001
    RETURN
C
    END
C
C#################################################################
C
    SUBROUTINE CUBIC(A,B,C,D,X1,X2,X3,SKIP)
    INTEGER SKIP
C
C   X1, X2 and X3 are the three roots of the cubic equation
C   A.X**3 + B.X**2 + C.X + D = 0
C   If no real values are found, for the given data (A,B,C,D),
C   SKIP is set to 1.
C
    SKIP = 0
    P = (3.0*A*C - B*B)/(9.0*A*A)
    Q = B**3/(27.0*A**3) - B*C/(6.0*A*A) + D/(2.0*A)
    IF(Q) 1,2,3
  1 EPS = 1.0
    GO TO 4
  3 EPS = -1.0
    GO TO 4
  2 IF(P.LT.0.0) GO TO 5
    Y1 = 0.0
    Y2 = (3.0*P)**0.5
    Y3 = -Y2
    GO TO 6
C
  4 IF(P.GT.0.0) GO TO 5
    DISC = P**3 + Q**2
    IF(DISC.GT.0.0) GO TO 5
    R = EPS*(-P)**0.5
    A1 = Q/R**3
    PHI = ACOS(A1)
    A2 = PHI/3.0
    A3 = (3.14159 - PHI)/3.0
    A4 = (3.14159 + PHI)/3.0
    Y1 = -2.0*R*COS(A2)
    Y2 = 2.0*R*COS(A3)
```

```
      Y3 = 2.0*R*COS(A4)
C
    6 COR = B/(3.0*A)
      X1 = Y1 - COR
      X2 = Y2 - COR
      X3 = Y3 - COR
      RETURN
C
    5 SKIP = 1
      X1 = 0.0
      X2 = 0.0
      X3 = 0.0
      RETURN
      END
C
C##############################################################
C
      SUBROUTINE RECON(NX,NY,DUR,NDELT,X,MODEL,P,DUH,PY)
      IMPLICIT REAL(A-Z)
      REAL X(150), DUR
     REAL S(1000),SDUH(1000),DUH(150),PY(150)
      REAL P(5,30)
      INTEGER I,J,NX,NY,NDELT,NINT,L,K,MAX0,MIN0,MODEL
C
C          CALCULATE S-CURVE BY TRAPEZOIDAL RULE
C
C    The S curve is calculated and stored in the array S at
C    intervals DT = DUR/NDELT. By increasing NDELT the resolution
C    of the S curve is increased, DT becoming smaller. NINT = NY *
C    NDELT contains the number of elements in the array S. The
C    first ordinate of every S curve is S(1) = 0.0 at time t = 0.
C    The S curve is built up using the trapezoidal rule. The
C    ordinates of the model IUH are supplied by a function
C    sub-program (IUH) to the temporary switched stores F1 and F2.
C    The arguments of IUH, namely, (T, MODEL, P) specify the point
C    in time T, the MODEL number, and its parameters P.
C
            NINT = NY*NDELT
            DT = DUR/NDELT
            S(1) = 0.0
                  F1 = IUH(0.0,MODEL,P)
      DO 14 I = 1,NINT
                  T = I*DT
                  F2 = IUH(T,MODEL,P)
  116             S(I+1) = S(I) + DT*(F1 + F2)/2.0
   14             F1 = F2
C
C      FIND UNIT PULSE RESPONSE FROM S-CURVE AT INTERVALS OF DT.
C
C    The unit pulse response is found in statement 15 as SDUH by
C    shifting the S curve forward in time by NDELT, subtracting
C    it from its original form and dividing by DUR = 1, the
C    duration of the input pulse.
```

```
C    Note that J runs backwards from J = NINT - 1 + 2 to
C    J = NINT - L + 2 = NINT - (NINT-NDELT +1) + 2 = NDELT + 1
C    i.e. backwards from NINT + 1 to NDELT + 1. The + 1 arises
C    because the array S is shifted by 1. S(1) is S at t = 0.
C    S(NINT + 1) is defined in statement 116. The second loop on I,
C    terminating on statement 20, completes the pulse response
C    simply by transferring S(I) to SDUH(I) for all I from 1 to
C    NDELT. There is no subtraction for these ordinates.
C
            L = NINT - NDELT + 1
       DO 15 I = 1,L
            J = NINT - I + 2
   15 SDUH(J) = (S(J) - S(J-NDELT))/DUR
C
       DO 20 I = 1,NDELT
   20 SDUH(I) = S(I)
C
C      FIND DUH AT INTERVALS OF DUR(=1) BY THE TRAPEZOIDAL RULE
C
C    SDUH is defined at NINT+1 points and contains NINT trapezia
C    between them.
C    These are accumulated in groups of NDELT and stored in DUH
C    at NY points.
C    The counter I in the DO loop to statement 16 runs over the
C    ordinates of SDUH, while the counter L runs more slowly
C    over DUH from 1 to NY = NINT/NDELT. The index J cycles
C    from 1 to NDELT repeatedly. At the IF statement,
C    when J is less than NDELT, J is increased by one,
C    and the next trapezium, number J, equal to
C    DT*(SDUH(I) + SDUH(I+1))/2.0 is added to those
C    already in DUH(L) in statement 16.
C    When J=NDELT at the final trapezium of a group, J returns
C    to zero, L is incremented by one, and DUH(L) is given an
C    initial value of zero, and in 16 the first trapezium of the
C    next group is accumulated in DUH(L).
C    Before the start of the DO loop on I, we set J=0 at the
C    start of the first (set L=1) group of SDUH values,
C    and DUH(L) is given an initial value of zero.
C
            J = 0
            L = 1
            DUH(L) = 0.0
C
       DO 16 I = 1,NINT
C      WRITE (6,112) S(I),SDUH(I)
  112 FORMAT(1H ,2(F10.7,2X))
       IF(J.LT.NDELT) GO TO 17
C      WRITE(6,113) L,DUH(L)
  113 FORMAT(1H ,24X,I4,F10.7)
       J = 0
       L = L + 1
       DUH(L) = 0.0
C
```

```
   17 J = J + 1
   16 DUH(L) = DUH(L) + DT*(SDUH(I) + SDUH(I+1))/2.0
C
C      ALTERNATIVE APPROACH...
C      DUH set equal to SDUH sampled at intervals of NDELT.
C If NDELT=1, J=I.
C
C    DO 216 I=1,NY
C          J=NDELT*(I-1)+1
C    DUH(I)=SDUH(J)
C 216 CONTINUE
C
C           CONVOLUTE X WITH DUH TO RECONSTITUTE Y AS PY
C
C    These five statements convolute DUH with X to yield a
C    predicted Y called PY. The index J runs from 1 to the
C    minimum of I and NX, the number of active X ordinates.
C    The sum of the arguments of DUH(I-J+1) and X(J)
C    in statement 18, is not I, as required by the definition of
C    convolution, but I+1, since all ordinates are shifted
C    by 1 in time. The elements of all arrays are counted
C    from (1,1,..).
C
     DO 18 I = 1,NY
           PY(I) = 0.0
           K = MINO(I,NX)
     DO 18 J = 1,K
  18 PY(I) = PY(I) + DUH(I-J+1)*X(J)
C
C    The RETURN transfers control back to the PICOMO sub-program,
C    i.e. to the statement immediately following the CALL
C    RECON statement.
C
     RETURN
     END
C
C##################################################################
C
     REAL FUNCTION IUH(T,MODEL,P)
     IMPLICIT REAL (A-Z)
     INTEGER MODEL, MOD
     REAL P(5,30)
C
     GO TO (1,2,3,4,5,6,7,8,9,10,11,12,13,14,15),MODEL
     MOD = MODEL - 19
     GO TO (20,21),MOD
C
   1 LRK = P(1,1)
     IUH = (1.0/LRK)*EXP(-T/LRK)
     RETURN
C
   2 BTR = P(1,2)
     IUH = 0.0
```

```
                  IF(T.GE.BTR) RETURN
                  IF(T.LE.BTR/3.0) GO TO 100
                  IUH = -3.0*(T-BTR)/(BTR*BTR)
                  RETURN
            100  IUH = 6.0*T/(BTR*BTR)
                  RETURN
      C
              3  LR2K = P(1,3)
                  IUH = (T/LR2K)*EXP(-T/LR2K)/LR2K
                  RETURN
      C
              4  SCT = P(1,4)
                  DT = P(2,4)
                  H1 = SCT - DT/2.0
                  H2 = SCT + DT/2.0
                  IUH = 0.0
                  IF(T.GE.H1.AND.T.LE.H2) IUH = 1.0/DT
                  RETURN
      C
              5  RRT = P(1,5)
                  RRK = P(2,5)
                  H = EXP(-T/RRK)
                  IF (T.GT.RRT) GO TO 113
                  IUH = (1.0-H)/RRT
                  RETURN
            113  H1 = EXP((RRT-T)/RRK)
                  IUH = (H1-H)/RRT
                  RETURN
      C
              6  RTT = P(1,6)
                  RTK = P(2,6)
                  H = 4.0/(RTT*RTT)
                  H1 = RTK*EXP(-T/RTK)
                  IF (T.GT.RTT/2.0) GO TO 111
                  IUH = H*(T-RTK+H1)
                  RETURN
            111  H2 = EXP(RTT/(2.0*RTK))
                  IF (T.GT.RTT) GO TO 112
                  IUH = H*(RTK-T+RTT-2.0*H1*H2+H1)
                  RETURN
            112  H3 = EXP(RTT/RTK)
                  IUH = H*(1.0-2.0/H2+1.0/H3)*H3*H1
                  RETURN
      C
              7  CRT = P(1,7)
                  CRK = P(2,7)
                  IUH = 0.0
                  IF (T.LT.CRT) RETURN
                  IUH = (1.0/CRK)*EXP((CRT-T)/CRK)
                  RETURN
      C
              8  CLIK = P(1,8)
                  BETA = P(2,8)
```

```
      H = EXP(-T/CLIK)
      IUH = H*(BETA*T/(CLIK*CLIK) + (1.0-BETA)/CLIK)
      RETURN
C
   9 LRNK = P(1,9)
      LRNN = P(2,9)
      H = T/LRNK
C     IUH = (H**(LRNN-1.0))*EXP(-H)/(LRNK*GAMMA(LRNN))
      RETURN
C
  10 C2RK1 = P(1,10)
      C2RK2 = P(2,10)
      IUH = EXP(-T/C2RK1) - EXP(-T/C2RK2)
      IUH = IUH/(C2RK1 - C2RK2)
      RETURN
C
  11 LCNK = P(1,11)
      LCNN = P(2,11)
      LCNT = P(3,11)
      IUH = 0.0
      IF(T.LT.LCNT) RETURN
      H = (T-LCNT)/LCNK
C     IUH = (H**(LCNN-1.0))*EXP(-H)/(LCNK*GAMMA(LCNN))
      RETURN
C
  12 BTR = P(2,12)
      A = P(1,12)
      IUH = 0.0
      IF(T.GE.BTR) RETURN
      IF(T.LE.A*BTR) GO TO 114
      IUH = -2.0*(T-BTR)/(BTR*BTR*(1.0-A))
      RETURN
 114 IUH = 2.0*T/(A*BTR*BTR)
      RETURN
C
  13 BR = P(1,13)
      IUH = 0.0
      IF(T.GT.BR) RETURN
      IUH = 1.0/BR
      RETURN
C
  14 A = P(1,14)
      IUH = 0.0
      IF(T.EQ.0.0) RETURN
      IUH = A/(3.14159*T**3)**0.5
      IUH = IUH*EXP(-A**2/T)
      RETURN
C
  15 A = P(1,15)
      B = P(2,15)
      IUH = 0.0
      IF(T.EQ.0.0) RETURN
      IUH = A/(3.14159*T**3)**0.5
```

```
            IUH = IUH*EXP(-(A-B*T)**2/T)
            RETURN
    C
       20 K1 = P(1,20)
          K2 = P(2,20)
          K3 = P(3,20)
          P(4,20) = 4.0
          H1 = K1*EXP(-T/K1)
          H2 = K2*EXP(-T/K2)
          H3 = K3*EXP(-T/K3)
          IUH = H1/((K1-K2)*(K1-K3)) + H2/((K2-K1)*(K2-K3))
        1                    +H3/((K3-K1)*(K3-K2))
          RETURN
    C
       21 BETA = P(1,21)
          K1 = P(2,21)
          K2 = P(3,21)
          H1 = EXP(-T/K1)
          H2 = EXP(-T/K2)
          IUH = (1.0 - BETA)*H2/K2
        1 +BETA*(H1-H2)/(K1-K2)
          RETURN
          END}
```

APPENDIX B

Inverse Problems are Ill-Posed

In this appendix we return to a number of questions in the earlier part of the book. We recall the special case of linear systems with finite memory and remark that the output is composed of two parts each described by a convolution integral differing only in their limits of integration. The first describes what may be called the fully developed response at all times greater than the memory length when the system has forgotten its initial state. The latter is usually taken as the fully relaxed state. The second describes the transient response when the system still remembers its initial state. When the memory is infinite, only the transient response occurs.

In the second section of this appendix we present the inverse problems of identification and detection, both of which are problems of deconvolution, as a problem in functional analysis. The concept of distance between two functions is central to this discussion. We quote standard results from the theory Euclidean function spaces showing that the chosen definition of distance is in agreement with our intuitive notions of distance.

Using this apparatus, we show that deconvolution is an ill-posed problem i.e. closer and closer fitting of the output of a linear system with finite memory does not guarantee closer and closer identification of the impulse response. The conclusion from this is the following. The search for the impulse response must be constrained to a sub-set of all possible response functions so that the search problem is well posed. This was the subject matter of Chapters 4 and 5.

B.1 LINEAR SYSTEMS WITH FINITE MEMORY

In Chapter 1 we presented the convolution integral for a linear time-invariant system with a finite memory M as equation (1.14),

$$y(t) = \int_{\tau=[t-M]}^{t} x(\tau)h(t-\tau)d\tau \tag{B.1a}$$

where the input, output and impulse response functions are $x(t), y(t)$ and $h(t)$ respectively. We now restrict these functions to continuous functions of time t on a closed interval $[0, T]$.

The impulse response function with finite memory $M < T$ is zero outside the time interval $[0, M]$

$$h(t) = 0 \quad \text{when } t \notin [0, M] \tag{B.1b}$$

For any value of t in the interval $[0,T]$ the convolution integral (B.1a) is formed from all $x.h$ product pairs whose arguments sum to t i.e. all pairs between $x(t - M)h(M)$ at the lower limit of integration and $x(t)h(0)$ at the upper limit. Since the input function may not have a negative argument, we wrote the lower limit of integration as $[t - M] = max(0, t - M)$ giving the value zero for $t < M$, and the value $t - M$ for $t > M$. The system is said to be *fully relaxed* prior to the time origin.

In other words, we may express (B.1a) as two separate cases: the *fully developed response* when the initial condition $x(t) = 0$, $t < 0$ has been "forgotten" at times $t > M$

$$y(t) = \int_{\tau=t-M}^{t} x(\tau)h(t - \tau)\mathrm{d}\tau, \quad M \leq t \leq T \tag{B.1c}$$

and the *transient response* from the fully relaxed initial condition $x(t) = 0$, $t < 0$ is at times $t < M$

$$y(t) = \int_{\tau=0}^{t} x(\tau)h(t - \tau)\mathrm{d}\tau, \quad 0 \leq t \leq M \tag{B.1d}$$

Continuity of output requires a common value of $y(M)$ in both cases. This distinction between fully developed and transient responses arises from the assumed finite memory of the system. When the memory is infinite, there is no fully developed response and the integral in (B.1d) applies to all values of t. The distinction does not depend on the nature of the input. It applies when the input is an isolated event, or a series of events.

An example will provide further clarification. Consider as input the unit step function at $t = 0$. In this case the transient response is the S-curve of duration M, given by the integral (B.1d). The fully developed response given by integral (B.1c) is the steady state response: $y(t) = x(t) = 1, t > M$, if, and only if, the system satisfies a conservation law i.e. the "area" under $h(t)$ is unity.

B.2 INVERSE PROBLEMS AND EUCLIDEAN FUNCTION SPACES

The *identification problem* is an *inverse problem* where we search for the best estimate of the impulse response $h(t)$ given an input/output pair of functions $x(t)$ and $y(t)$. The *detection problem* is also an inverse problem where we search for the best estimate of the input function $x(t)$ given an impulse–response/output pair of functions $h(t)$ and $y(t)$. The following discussion applies to both inverse problems, but is presented in terms of the first.

We begin by defining the "distance" between two functions in a suitable *function space*. We will use this concept to discuss questions concerning the estimation or approximation of functions in the inverse problem. When there are many possible approximate functions, we may also search for the best approximate function with the property that it lies closest to the given function in the function space.

We choose as our function space, the space F of all continuous functions, defined on a closed interval of time $[0, T]$ (See for example Kreider et al., 1966, Chapter 7: Euclidean Spaces). If $c(t)$ and $d(t)$, $t\varepsilon[0, T]$ are a pair of functions in F, we may define their *inner product* as

$$\langle c, d \rangle_F = \int_0^T c(t)d(t)\mathrm{d}t \tag{B.2}$$

Using the inner product we define the length or *norm of any function $e(t)$* in F, as

$$\|e\|_F = \sqrt{\langle e, e \rangle_F} = \sqrt{\int_0^T e^2(t)\mathrm{d}t} \tag{B.3}$$

which is always a non-negative quantity. Other definitions of length are possible. Finally, we define the *distance between two functions $a(t)$ and $b(t)$* in F as the length, or norm, of their difference $e(t) = a(t) - b(t)$

$$d_F(a, b) = \|a - b\|_F = \sqrt{\int_0^T (a(t) - b(t))^2\mathrm{d}t} \tag{B.4}$$

It can be shown (Kreider et al., 1966, Chapter 7) that $d_F(a, b)$ is a real number satisfying our intuitive notions of distance

$d_F(a, b) \geq 0$ (*Non-negativity*)

$d_F(a, b) = 0$ *if and only if $a = b$*   (*Coincidence*)

$d_F(a, b) = d_F(b, a)$ (*Symmetry*)

$d_F(a, b) + d_F(b, c) \geq d_F(a, c)$ *for any three functions a,b,c*   (B.5)
(*Triangle inequality*)

$d_F(\xi a, \xi b) = |\xi| \cdot d_F(a, b)$ *for any real number ξ*   (*Scaling*)

$d_F(a + c, b + c) = d_F(a, b)$ *for any three functions a,b,c*
(*Shifting*)

We now define two different function spaces; the first is the space H of continuous impulse response functions of finite memory defined on the closed interval $[0, M]$; the second is the space Y of continuous output functions defined on the closed interval $[0, T]$.

Let $h(t)$ be the "true" impulse response when the output function is $y(t)$ i.e.

$$y(t) = h(t) * x(t) \tag{B.6}$$

Let $k(t)$ be the estimated impulse response when the output contains errors i.e. $k(t)$ is the solution of the inverse problem of de-convolution

$$z(t) = k(t) * x(t) \tag{B.7}$$

where $z(t)$ is the erroneous output.

In both function spaces H and Y, we define a distance between any function in the space and a reference function in the same space. We define $d_Y(z, y)$ to be the distance in Y between the output functions z and y. We define $d_H(k, h)$ to be the distance in H between the impulse response function k which satisfies $z(t) = k(t) * x(t)$ and the impulse response h which satisfies $y(t) = h(t) * x(t)$. The functions z and y correspond to k and h, respectively. The functions y and h are the reference functions in their respective spaces. When there are no errors, then $z = y$, $d_Y(z, y) = 0$, which in turn implies $k = h$ and $d_H(k, h) = 0$.

An inverse problem is said to be *well-posed* whenever there is a search procedure in H space with the property

$$d_Y(z, y) \xrightarrow[H]{} 0 \Rightarrow d_H(k, h) \to 0 \tag{B.8}$$

The first arrow indicates any search procedure in H applied to $k(t)$ which systematically reduces the distance $d_Y(z, y)$ in Y space — *fitting the given output y better and better*. The second arrow indicates that the search procedure in H has the property that there is a corresponding reduction in $d_H(k, h)$ in H space and *k approaches the unknown h function as the search proceeds*. In other words, the problem is well-posed, whenever fitting the output data, implies fitting the impulse response as well. An inverse problem is said to be *ill-posed* when this is not the case. The following argument due to Napiorkowski (Napiorkowski and Strupczewski, 1984; Napiorkowski, J.J., 1988) shows that de-convolution is, in general, an ill-posed problem.

B.3 DECONVOLUTION IS AN ILL-POSED PROBLEM

The argument begins with a sinusoidal perturbation of the impulse response, thereby avoiding the direct solution of the inverse problem of deconvolution. Let $k(t)$ be a sinusoidal perturbation of $h(t)$ that depends on two parameters, N (amplitude) and ω (frequency)

$$k(t) = h(t) + N \sin(\omega t) \tag{B.9}$$

In order to examine whether (B.8) is true, or not, in the case of linear systems with finite memory, we use (B.9) to find expressions for both

$d_Y(z, y)$ and $d_H(k, h)$. The square of the distance between the functions $k(t)$ and $h(t)$ in H space is

$$
\begin{aligned}
d_H^2(k, h) &= ||k - h||_H^2 = \int_0^M [k(\tau) - h(\tau)]^2 d\tau \\
&= N^2 \int_0^M \sin^2(\omega\tau) d\tau \\
&= \frac{N^2}{2} \left(M - \frac{1}{\omega} \sin(\omega M) \cos(\omega M) \right) \\
&= \frac{N^2 M}{2} \quad as\ \omega \to \infty \qquad (B.10)
\end{aligned}
$$

The search procedure referred to in (B.8) is parameterised on N and ω. Allowing N and ω to vary, defines a path in H space.

The square of the distance $d_Y(z, y)$ between the functions $z(t)$ and $y(t)$ in Y space is much more difficult to evaluate than $d_H(k, h)$. Substituting (B.9) in the convolution integral (B.1c) for a system with finite memory M we find the *fully developed response*

$$
\begin{aligned}
z_{fdr}(t) &= k(t) * x(t) \\
&= h(t) * x(t) + N \sin(\omega \cdot t) * x(t) \\
&= y(t) + \int_{\tau=t-M}^t N \sin \omega(t - \tau) x(\tau) d\tau \quad M \le t \le T \quad (B.11a)
\end{aligned}
$$

and substituting in (B.1d) yields the *transient response*

$$
z_{tr}(t) = y(t) + \int_{\tau=0}^t N \sin \omega(t - \tau) x(\tau) d\tau \quad 0 \le t \le M \qquad (B.11b)
$$

The square of the distance between z and y in the output function space Y is

$$
\begin{aligned}
d_Y^2(z, y) &= ||z - y||_Y^2 \\
&= \int_{t=0}^T (z(t) - y(t))^2 dt \\
&= \int_{t=0}^M (z_{tr}(t) - y(t))^2 dt + \int_{t=M}^T (z_{fdr}(t) - y(t))^2 dt \\
&= \int_{t=0}^M \left(\int_{\tau=0}^t N \sin[\omega(t - \tau)] x(\tau) d\tau \right)^2 dt \\
&\quad + \int_{t=M}^T \left(\int_{\tau=t-M}^M N \sin[\omega(t - \tau)] x(\tau) d\tau \right)^2 dt \qquad (B.12a)
\end{aligned}
$$

An explicit expression for (B.12a) is not possible in the case of the general input function $x(t)$. It is sufficient for our purposes to show that the square root of (B.12a), which is inherently non-negative, has the asymptotic property

$$dy(z,y) \to 0 \quad \text{as } \omega \to \infty \tag{B.12b}$$

for any value of N. The argument proceeds from the simple to the more complicated. We begin with the simplest case of a step input at the origin. We then shift the step forward in time and examine the distance between the corresponding functions z and y in Y space. The argument ends with the asymptotic behaviour of any bounded input, which is approximated by a superposition of shifted step functions.

Take the step function at the origin as the input $x(t) = s, t \geq 0$. Consequently, $x = s$ in both inner integrals in expression (B.12a). The first inner integral, namely, the error in the S-curve transient response, is solved using the change of variable $u = t - \tau, du = -d\tau$, with the limits of integration $u = 0$ when $\tau = t$, and $u = t$ when $\tau = 0$, for all values of t in the closed interval $0 \leq t \leq M$,

$$\int_0^t N \sin \omega(t - \tau)s \, d\tau = \int_0^t sN \sin(\omega u) du$$

$$= \frac{sN}{\omega}(1 - \cos(\omega t)) \quad 0 \leq t \leq M \tag{B.13a}$$

Using the same change of variable, $u = t - \tau, du = -d\tau$, with the limits of integration $u = 0$ when $\tau = t$, and $u = M$ when $\tau = t - M$, we discover that the second inner integral is independent of time i.e. the error in this fully developed response is a constant for all values of t in the closed interval $M \leq t \leq T$.

$$\int_{t-M}^t N \sin \omega(t - \tau)s \, d\tau = \int_0^M sN \sin(\omega u) du$$

$$= \frac{s \cdot N}{\omega}(1 - \cos(\omega M)) \quad M \leq t \leq T \tag{B.13b}$$

Substituting in (B.12a) we find

$$d_Y^2(z,y) = \int_{t=0}^M (z_{tr}(t) - y(t))^2 dt + \int_{t=M}^T (z_{fdr}(t) - y(t))^2 dt$$

$$= \left(\frac{sN}{\omega}\right)^2 \left(\int_{t=0}^M (1 - \cos(\omega t))^2 dt \right.$$

$$\left. + \int_{t=M}^M (1 - \cos(\omega M))^2 dt \right) \tag{B.14a}$$

The first integral in (B.14a), is

$$\int_{t=0}^M (1 - \cos(\omega t))^2 dt = \frac{3}{2}M - \frac{2\sin(\omega M)}{\omega} + \frac{\sin(2\omega M)}{4\omega} \tag{B.14b}$$

The second integral in (B.14a), is

$$\int_{t=M}^{T} (1 - \cos(\omega M))^2 dt = (1 - \cos(\omega M))^2 (T - M) \tag{B.14c}$$

Combining (B.14a, b and c) gives

$$
\begin{aligned}
d_Y^2(z,y) &= \|z - y\|_Y^2 \\
&= \int_{t=0}^{T} [z(t) - y(t)]^2 dt \\
&= \left(\frac{sN}{\omega}\right)^2 \left[\begin{array}{l} \dfrac{3}{2}M - \dfrac{2 \cdot \sin(\omega M)}{\omega} + \dfrac{\sin(2\omega M)}{4 \cdot \omega} \\ +(T - M)(1 - \cos(\omega M))^2 \end{array} \right]
\end{aligned} \tag{B.14d}
$$

Clearly the distance between the functions z and y tends to zero as ω tends to infinity for any value of N.

Now consider a step-input of magnitude s_D shifted by a duration D

$$x(t) = s_D H(t - D), \quad 0 \le D \le T - M \tag{B.15}$$

where H is the Heaviside or unit step function. From the definitions we have

$$
\begin{aligned}
z(t) &= k(t) * x(t) \\
&= h(t) * x(t) + N \sin(\omega t) * x(t) \\
&= y_D(t) + N \sin(\omega t) * s_D H(t - D)
\end{aligned} \tag{B.16a}
$$

and

$$
\begin{aligned}
d_Y^2(z, y_D) &= \int_{t=0}^{T} [z(t) - y_D(t)]^2 dt \\
&= \int_{t=0}^{T} [N \sin(\omega t) * s_D H(t - D)]^2 dt
\end{aligned} \tag{B.16b}
$$

The input and output functions are both zero for $0 < t < D$. Hence, the *transient response* in this case is defined on the interval $D < t < M + D$, with $M + D < T$, and the *fully* developed response is defined on the interval $M + D < t < T$. The convolution under the integral-square operator in (B.16b) now has three parts rather than two. The expressions for these are the results in (B.13a and b) shifted forward in time by D.

$$N \sin(\omega . t) * s_D H(t - D)$$

$$
= \begin{cases} 0 & 0 \le t \le D & \text{(B.17a)} \\ \left(\dfrac{N s_D}{\omega}\right)(1 - \cos \omega(t - D)) & D \le t \le M + D & \text{(B.17b)} \\ \left(\dfrac{N s_D}{\omega}\right)(1 - \cos \omega M) & M + D \le t \le T & \text{(B.17c)} \end{cases}
$$

Squaring these functions and integrating over their respective domains gives

$$d_Y^2(z, y_D) = \left(\frac{s_D N}{\omega}\right)^2 \left(\int_{t=D}^{M+D} (1 - \cos\omega(t - D))^2 dt \right.$$

$$\left. + (1 - \cos\omega M)^2 \int_{t=M+D}^{T} dt\right) \tag{B.18a}$$

Using the change of variable $u = t - D$, $du = dt$ with the limits $t = D$, $u = 0$ and $t = M + D$, $u = M$, equation (B.18a) becomes almost identical to (B.14d) namely

$$d_Y^2(z, y_D) = \left(\frac{s_D N}{\omega}\right)^2 \left(\int_{u=0}^{M} (1 - \cos\omega \cdot u)^2 du\right.$$

$$\left. + (1 - \cos\omega M)^2 \int_{t=M+D}^{T} dt\right)$$

$$= \left(\frac{s_D N}{\omega}\right)^2 \left[\begin{array}{l} \frac{3}{2}M - \frac{2\sin(\omega M)}{\omega} + \frac{\sin(2\omega M)}{4\omega} \\ +(T - M - D)(1 - \cos(\omega M))^2 \end{array}\right] \tag{B.18b}$$

with exact equality when $D = 0$. Hence, in the case of the shifted step function, we may also conclude that the distance between the functions z and y tends to zero as ω tends to infinity for any fixed value of N.

Finally, we consider the general case of a bounded input function $x(t)$ approximated to any specified accuracy using a superposition of shifted step functions

$$x(t) = \sum_G s_D H(t - D), \quad G = \{D | D \in R \text{ and } 0 \leq D \leq T - M\},$$

$$s_D \in R \tag{B.19}$$

s_D is the step size at time D; s_D may be a positive or negative real number in the set R; H is the unit step or Heaviside function located at each $t = D$. G is the set of times at which the steps occur. The corresponding output is

$$z(t) = k(t) * \sum_G s_D H(t - D)$$

$$= h(t) * \sum_G s_D H(t - D) + N\sin(\omega t) * \sum_G s_D H(t - D)$$

$$= y(t) + \sum_G s_D N \sin(\omega t) * H(t - D) \tag{B.20}$$

The square of the distance between z and y is

$$d_Y^2(z, y) = \int_{t=0}^{T} [z(t) - y(t)]^2 dt$$

$$= \int_{t=0}^{M} [s_{D1} N \sin(\omega t) * H(t - D_1)$$

$$+ s_{D2} N \sin(\omega t) * H(t - D_2) + \cdots]^2 dt \qquad (B.21)$$

Each convolution in (B.21) has three parts corresponding to those in (B.17a, b and c). Call each part, a, b and c, in the case of $H(t - D_1)$, A, B and C, in the case of $H(t - D_2)$, α, β and γ, in the case of $H(t - D_3)$, and so on. When we rank the D_s, $D_1 < D_2 < D_3 < \ldots$ the location of M in the ranked list gives rise to many possible cases. When there are only two D_s, $D_1 < D_2 < M$, the domain of integration in (B.21) has five parts each with a different integrand:

- In the interval 0 to D_1 the integrand is $(a + A)^2 = 0$
- In the interval D_1 to D_2 the integrand is $(b + A)^2 = b^2$
- In the interval D_2 to $D_1 + M$ the integrand is $(b + B)^2 = b^2 + 2bB + B^2$
- In the interval $D_1 + M$ to $D_2 + M$ the integrand is $(c + B)^2 = c^2 + 2cB + B^2$
- In the interval $D_2 + M$ to T the integrand is $(c + C)^2 = c^2 + 2cC + C^2$

In each case, the integrand contains the common factor $(N/\omega)^2$ multiplied by a sum of products of pairs of time-shifted or constant cosine functions. When there are three D_s, the integrands are cyclic variations on $(b + B + \beta)^2 = b^2 + B^2 + \beta^2 + 2bB + 2b\beta + 2B\beta$. Each integrand contains the same common factor $(N/\omega)^2$ multiplied by a sum of products of pairs of time-shifted or constant cosine functions, and so on. The integration of these pairs of cosine functions may place ω in the denominator, as in (B.18b), but not in the numerator. Hence

$$d_Y^2(z, y) = \left(\frac{N}{\omega}\right)^2 f\left(T, M, s_{D1}, D_1, s_{D2}, D_2, \ldots, \sin -, \right.$$

$$\left. \sin^2 -, \cos^2 -, \ldots; \frac{1}{\omega}\right) \qquad (B.22)$$

Consequently, for any bounded input function $x(t)$ on the interval $[0, T]$ with T sufficiently large, the square of the distance between the two corresponding output functions z and y tends to zero as ω tends to infinity for any fixed value of N.

On the other hand, in the H function space, we have already established that

$$d_H^2(k, h) = \frac{N^2}{2}\left(M - \frac{1}{\omega}\sin(\omega M)\cos(\omega M)\right)$$

$$= \frac{N^2 M}{2} \quad \text{as } \omega \to \infty \qquad (B.23)$$

Hence, the distance between k and h in H space can be made arbitrarily large

$$d_H(k,h) \to \infty \quad \text{as } \omega \to \infty \quad \text{and} \quad N \to \infty \tag{B.24a}$$

while the distance between z and y in Y space can be made arbitrarily small.

$$d_Y(z,y) \to 0 \quad \text{as } \omega \to \infty \quad \text{for any} \quad N \tag{B.24b}$$

Any specified distance between z and y, however small, can be achieved by certain values of the ratios N/ω and $1/\omega$ in (B.23). By forming the desired ratios from larger and larger values of both N and ω, we can make the corresponding distance between k and h indefinitely large. These contrary asymptotic limits can be regarded as the outcome of a search procedure in *H space* parameterised on N and ω. An inverse problem is said to be ill-posed when there is a search procedure in H that drives $d_Y(z,y)$ to zero while $d_H(k,h)$ grows larger and larger. The essential conclusion is that very good fitting of the predicted output z to the measured data y may be achieved with a completely misleading impulse response $k(t)$, and extends the corresponding experimental and theoretical results of Chapter 4 to continuous time. The demonstration shows that general search procedures in the function space H cannot be expected to have the property that $d_Y(z,y) \to 0 \Rightarrow d_H(k,h) \to 0$. See Napiorkowski and Strupczewski (1984) and Napiorkowski (1986, 1988) for the non-linear case and for other search techniques in H. See also Tikhonov (1963, 1965), Kuchment (1967) and Tikhonov and Arsenin (1974) for earlier discussion.

To overcome the ill-posed nature of the problem we may introduce additional mathematical and physical principles to restrict H to a subset H_S within which the inverse problem may be *well-posed* i.e. where the search procedure in H_S has the property that $d_Y(z,y) \to 0$ implies $d_H(k,h) \to 0$. We saw in Chapter 4 several examples of such principles in discrete time, such as conservation of water volume, non-negativity of the impulse response function and absence of oscillations. But even these may be insufficient to guarantee that the inverse problem of de-convolution is always well-posed in hydrology.

For example, we can impose conservation of water volume by requiring (B.9) to satisfy

$$\int_{t=0}^{M} k(t)\mathrm{d}t = \int_{t=0}^{M} h(t)\mathrm{d}t = 1 \Leftrightarrow \int_{t=0}^{M} N\sin(\omega t)\mathrm{d}t = 0 \tag{B.25}$$

Consequently, the search procedure in H space is no longer parameterised on the continuous variable ω, but is restricted to a discrete set of values, enumerated on n, which satisfy

$$\omega M = 2\pi n, \quad n = 1, 2, 3 \ldots \tag{B.26}$$

thereby ensuring that (B.25) is also satisfied. In other words, the sinusoidal perturbation in (B.9) is now an integral number of sine waves that fit exactly into the interval $[0, M]$. Substituting (B.26) in (B.14d) or (B.18b) we find for any step function input

$$d_Y^2(z, y) = \frac{3M}{8\pi^2} \left(\frac{sNM}{n}\right)^2$$ (B.27a)

Substituting (B.26) in (B.10) we find

$$d_H^2(k, h) = \frac{1}{2}MN^2$$ (B.27b)

A search procedure ordered on the discrete set of values of ω in (B.26) and on N, in the same manner as before, allows us to conclude that the distance between k and h in H space can be made arbitrarily large, while the distance between z and y in Y space can be made arbitrarily small. Therefore, constraining the search in H space to functions $k(t)$ that conserve water volume is not sufficient to ensure that the inverse problem is always well-posed.

We saw in the case of the water-conserving conceptual models (*COMO*) in Chapter 5, how simple assumptions concerning the distribution of lateral inflow, translation in space and the presence of storage lead to a great reduction in the H space to a sub-set we may call H_{COMO}. The inverse problem for conceptual models in H_{COMO} is stable to small changes in the output data. However, the "true" impulse response may not be present in H_{COMO}. Reducing the distance $d_Y(z, y)$ between z and y, may be accompanied by a reduction in the distance $d(k, h)$ between k and h in H_{COMO}; but $d(k, h)$ cannot be driven to zero, except in very unusual circumstances. The question then becomes: When is $d(k, h)$ close enough to zero? The program PICOMO described in Appendix A is a tool for exploring these issues.

REFERENCES

Kreider, D.L., Kuller, R.G., Ostberg, D.R. and Perkins, F.W. (1966). *An Introduction to Linear Analysis*. Addison-Wesley, Reading, Mass.

Kuchment, L.S. (1967). Solution of inverse problems for linear flow models. *Sov. Hydrol. Sel. Pap.* 2, pp. 194–199.

Napiorkowski, J.J. (1986). Application of Volterra series to modelling of rainfall-runoff systems and flow in open channels. *Hydrol. Sci. J.*, 31, pp. 187–193.

Napiorkowski, J.J. (1988). *Zastosowania Szeregu Volterry w Hydrologii Dynamicznej. Prace Habilitacyjne.* Polska Akademia Nauk, Instytut Geofizyki, Warsaw.

Napiorkowski, J.J. and Strupczewski, W.G. (1984). Problems involved in the identification of kernels of Volterra series. *Acta Geophys. Polon.*, 32, pp. 375–391.

Tikhonov, A.N. (1963). O Resheniy Nekorrektno Postavlennykh Zadach i Metode Regularizatsyi (On the Solution of Ill-posed Problems and the Regularization Method). *DAN*, 151.

Tikhonov, A.N. (1965). Improperly posed problems and a stable method for their solution. *Sov. Math.*, 6, pp. 988–991.

Tikhonov, A.N. and Arsenin, V.Y. (1974). *Metody Resheniya Nekorrektnykh Zadach (Methods of Solving Ill-posed Problems)*. Nauka, Moscow.

APPENDIX C

The Non-Linearity of the Unsaturated Zone

Chapter 2 presented a brief description of hydrological systems and was followed by four chapters devoted to the systems approach as applied to direct surface runoff. The purpose of this appendix is to provide the reader with a much more detailed treatment of one part of the hydrological cycle, namely, the hydrology of the unsaturated zone. This provides a bridge between Chapter 2 and Chapters 7 and 8. Chapter 7 deals with simple models of sub-surface flow, while Chapter 8 is concerned with non-linear deterministic models.

The material in this appendix covers the principal sources of non-linearity in the underlying physics of the unsaturated zone with the exception of hysteresis. We derive the non-linear equations and boundary conditions of the Philip–Richards model of isothermal movement of water in both the liquid and vapour phase in a vertical column of soil. This model encompasses the processes of infiltration, evaporation, redistribution, capillary rise and drainage at the pedon scale of one meter. The dominant non-linearities arise from two pairs of switches expressed in the boundary conditions: an outer pair always switching between wet and dry periods, and an inner pair switching intermittently from atmosphere control to soil control of the surface fluxes. The model can be extended to include transpiration and hysteresis, but is not presented here.

Particular attention is paid to units, since at least three different systems are in use in the literature, a source of confusion to the beginner. The system of units in this appendix may be called the strict *SI* system in contrast to the nominally simpler engineering system of mixed units used in Chapter 7.

C.1 WATER MOVEMENT IN A SOIL COLUMN

Equations of motion and conservation of mass

The two equations that govern the vertical movement of water in an unsaturated rigid column of bare soil at the pedon scale are taken from soil-physics (see for example Feddes, 1971 or Campbell, 1985). These are

(1) a water balance, that states that water mass is conserved; and
(2) a dynamic equation, with liquid and vapour water flux proportional to the gradient in soil-water potential.

If z is a position coordinate measured vertically downwards — the opposite direction to that in equation (7.7) in Chapter 7 — from the soil surface and t is time, we may write these equations as follows

$$\rho_w \frac{\partial \theta}{\partial t} = -\frac{\partial (f_l + f_v)}{\partial z} \quad 0 \le \theta \le \theta_s < 1 \tag{C.1a}$$

$$f_l = -k_l \frac{\partial \psi}{\partial z} \tag{C.1b}$$

$$f_v = -k_v \frac{\partial \psi}{\partial z} \tag{C.1c}$$

where ρ_w is the *density of water* ($kg\,m^{-3}$); θ is the *volumetric moisture content* (m^3 of soil water per m^3 of rigid soil/void matrix — this is the moisture content c in the corresponding equations (7.8–7.10) in Chapter 7; θ_s is the *saturation* volumetric moisture content; f_l is the *flux density* of liquid water ($kg\,m^{-2}\,s^{-1}$) in the direction of positive z i.e. downwards, which is assumed to be proportional to the gradient in the *soil water potential* $\partial \psi / \partial z$ (Buckingham, 1907) — ϕ is water potential in Chapter 7 but has different units to ψ, see below; f_v is the flux density of water vapour ($kg\,m^{-2}\,s^{-1}$), which is also assumed to be proportional to the gradient in the soil water potential $\partial \psi / \partial z$ (Philip, 1955). The *liquid* and *vapour conductivities* k_l, and k_v, may be considered, in the first instance, to be functions of water content. Uppercase K is used in Chapter 7 for conductivity; but the units are different; see below.

In Chapter 7 the flux density is presented, not as a mass flux, but as a volumetric flux of cubic meters of water per square meter of vertical column of soil per second ($m^3\,m^{-2}\,s^{-1}$ or $m\,s^{-1}$) which has units of velocity. Consequently, equation (C.1b) corresponds to equation (7.7) of Chapter 7, but with a different set of units. Similarly, equation (7.10) is a dynamic water balance whereas the corresponding equation (C.1a) is a dynamic mass balance.

Soil-water potential

Different authors define the *potential energy of soil water* at the pedon scale, in one of three equivalent ways (Rose, 1966) as

- energy per unit mass, or
- energy per unit volume, or
- energy per unit weight.

The difference between these three alternative sets of units may be seen most easily by considering the change in potential energy of a mass m of water — occupying a volume V at a density of $\rho_w = m/V$ — when it is moved from one elevation to another in the gravitational field close to the Earth's surface. The necessary force is $mg = \rho_w V g$. The force is applied along any path connecting the initial and final vertical positions a distance z

apart. We assume there is no conversion of energy to heat through friction. Consequently, the work done is dissipationless work against the force of gravity and is equal to $mgz = (\rho_w V g)z$. The three alternative ways of representing the potential energy of this water as dissipationless work (a) per unit mass, (b) per unit volume, and (c) per unit weight are

$$\psi_{mass} = \frac{mgz}{\rho_w V} = \frac{\rho_w V gz}{\rho_w V} = gz \quad (\text{J kg}^{-1}) \tag{C.2a}$$

$$\psi_{volume} = \frac{mgz}{V} = \frac{\rho_w V gz}{V} = \rho_w gz \quad (\text{J m}^{-3} \quad \text{or} \quad \text{N m}^{-2}) \tag{C.2b}$$

$$\psi_{weight} = \frac{mgz}{\rho_w Vg} = \frac{\rho_w V gz}{\rho_w Vg} = z \quad (\text{m}) \tag{C.2c}$$

Their ratios are

$$\psi_{mass} : \psi_{volume} : \psi_{weight} = g : \rho_w g : 1 \tag{C.2d}$$

These may be used to convert from one form to the other (Feddes, 1971).

There are three matching sets of units for each of the three quantities: flux density of soil-water, soil-water conductivity, and gradient in the soil-water potential. This arises from the generalised *Darcy equation* formed from equations (C.1b and c)

$$f = f_l + f_v$$
$$= -(k_l + k_v)\frac{\partial \psi}{\partial z} \tag{C.3}$$
$$= -k\frac{\partial \psi}{\partial z}$$

The three sets of units, and their underlying physical dimensions, must be self-consistent with respect to equation (C.3). The following table shows how the units of k are defined to ensure consistency.

Term/Basis	Mass	Volume	Mixed volume and weight
Water flux density f	$\text{kg m}^{-2}\text{s}^{-1}$	$\text{m}^3\text{ m}^{-2}\text{s}^{-1}$ *or* m s^{-1}	$\text{m}^3\text{ m}^{-2}\text{s}^{-1}$ *or* m s^{-1}
Soil-water potential ψ	J kg^{-1}	J m^{-3} *or* N m^{-2} *or* Pa	m
Gradient in potential $\partial\psi/\partial z$	$\text{J kg}^{-1}\text{m}^{-1}$ *or* N kg^{-1} *or* $\text{m}^3\text{ m}^{-2}\text{s}^{-2}$ *or* m s^{-2}	Pa m^{-1} *or* $\text{kg m s}^{-2}\text{ m}^{-3}$ *or* $\text{kg m}^{-2}\text{s}^{-2}$	m m^{-1}
Hydraulic conductivity k	kg s m^{-3}	$\text{m}^3\text{ s kg}^{-1}$	m s^{-1}

The column headed "Mass" in the table shows (a) the water flux density measured as a mass flow rate per unit area perpendicular to the flow direction, and (b) the soil-water potential measured as energy per unit mass of water, both in strict *SI* units. Consequently, the gradient in soil-water potential has units of Joules per kilogram per meter, which is equivalent to Newtons per kilogram of soil-water, which is also its acceleration, in accordance with the definitions of the Joule and the Newton. For consistency, k must have the units $kg\,s\,m^{-3}$.

In the column headed "Volume", water flux density is measured as a volumetric flux per unit area and has units of velocity. From (C.2b) the soil-water potential, measured as energy per unit volume, is equivalent to a pressure and is quoted in the literature in a variety of units, such as kilo-Pascals, bar or an equivalent column of mercury. The last two entries in the second column show the corresponding *SI* units for (a) the gradient in soil-water potential, and (b) the hydraulic conductivity, so that the whole column is self-consistent.

Finally, the last column shows a mixed system based on (a) volume, rather than weight, for the water flux density, and (b) energy per unit weight as the measure of soil-water potential in equation (C.2c). The potential energy per unit weight of soil-water is measured by the height of an equivalent column of water, or so-called *pressure head*. The gradient in potential energy has units of meter per meter. Consequently, for consistency, the hydraulic conductivity must be measured in the same units as the water flux density, namely, cubic meters of water per square meter per second, or a water velocity. This is the simplest system of all and may be called the engineering system of units. This system is used in Chapter 7.

Component soil-water potentials

The potential ψ is assumed to be the sum of several *component potentials* measured with respect to a common *reference state*. The most important of these are the gravitational potential ψ_g and the matric potential ψ_m. Hence

$$\psi = \psi_m + \psi_g \quad \psi \leq 0 \tag{C.4a}$$

The *gravitational potential* is negative in equation (C.2a) when the position coordinate is taken as positive downwards from the surface. Hence

$$\psi_g = -g\,z \tag{C.4b}$$

The equation corresponding to (C.4a and b) in Chapter 7 (7.2 on page 133), is $\phi = h = p/\gamma + z$ i.e. "the potential ϕ will be equal to the *piezometric head h* i.e. the sum of the *pressure head* p/γ and the elevation z, where p is the pressure in the soil water, γ is the *weight density of the water*, and z is the elevation above a fixed horizontal datum". In equation (7.2) potential energy ϕ is represented as energy per unit weight with the weight density given by $\gamma = \rho_w g$. The corresponding gravitational potential is z, not $-gz$, in accordance with the engineering system of units in (C.2d) above. The sign is reversed because the vertical coordinate is positive upwards in Chapter 7, whereas it is positive downwards in this appendix; this confusion is reflected in the eclectic literature on soil-water.

The matric potential of soil water ψ_m ($J\,kg^{-1}$) is also negative and is a function of water content. Thermo-mechanical work must be done in order to return the soil water to the reference state, by removing it from the soil, whether by suction, drainage or evaporation. In Chapter 7 (page 134) "in order to avoid continual use of negative pressures, it is convenient and is customary in discussing unsaturated flow in porous media to use the negative of the pressure head and to describe this as soil suction (S_s)" i.e. $S_s = -p/\gamma$ in the engineering system of units. The logarithm to the base 10 of S_s expressed as centimeters of water in tension, is the pF scale of measurement, by analogy with the pH scale.

The relationship between water content and matric potential, the *soil-water characteristic* $\psi_m = \psi_m(\theta)$, is a constitutive property of the soil-air-water system and is determined by measurements on soil samples at the pedon scale. The inverse function $\theta = \theta(\psi_m)$ and its derivative, the *specific water capacity* $C(\theta) = d\theta/d\psi_m$, may be used to eliminate θ from equation (C.1a), from k_l, and k_v, yielding ψ_m as the dependent variable. θ may also be chosen as the dependent variable. This is the case in Chapter 7 and leads to equation (7.11). When the soil is layered, θ_s and k_l are discontinuous functions of depth. Consequently, the choice of ψ_m as the dependent variable is more general than θ, since ψ_m may be a continuous function of z when θ is not.

C.2 THE PHILIP–RICHARDS EQUATION

Combining equations (C.1, C.3 and C.4) and eliminating the fluxes, we find the equation for the unknown matric potential ψ_m

$$\rho_w C \frac{\partial \psi_m}{\partial t} = -\frac{\partial}{\partial z}\left(k\frac{\partial \psi_m}{\partial z} - kg\right) \tag{C.5}$$

where $k = k_l + k_v$. When $k_v = 0$ this is known as Richards equation (1931) of soil physics; the extension to include vapour transport is due to Philip (1955). This is a non-linear parabolic equation of the diffusion type with "concentration" dependent coefficients $C(\psi_m)$ and $k(\psi_m)$.

Extensions of this equation to include two or three space dimensions, hysteresis, the non-rigid matrices of soils, the presence of electrolytes and the effect of temperature gradients, are also possible. The above presentation is sufficient for this discussion.

C.3 POTENTIAL ENERGY AND THE PHASE PARTITIONING OF WATER

The potential energy, ψ, of water in a state S is equal to the notional work that must be performed by a given process, in order to convert water from a reference state S_0, to the state S. The reference state is free pure liquid water at $z = 0$ in isothermal equilibrium with its vapour at saturation in a standard atmosphere.

Liquid water in an unsaturated soil is not free, since it is bound to the soil matrix by capillary and adhesive forces. Furthermore, the concentration of water vapour in the soil-air, which is in equilibrium with liquid soil-water, has a value less than that in the reference atmosphere.

The work done, $\psi_h < 0$, in expanding isothermally the water vapour in the standard atmosphere in its reference state S_0, to achieve a lower humidity is

$$\psi_h = \frac{RT}{M_w} \ln h \quad 0 < h \le 1 \tag{C.6a}$$

assuming that atmospheric water vapour obeys the ideal gas and partial pressure laws. M_w is the molar mass of water (0.018 kg/mol), R the molar gas constant (8.314 J/mol K) and T the constant temperature in degrees Kelvin (293 K at 20°C).

On the other hand, $\psi_m < 0$, is the work done in transferring isothermally the liquid water from the state S_0 into an inert rigid porous soil. We may equate ψ_h and ψ_m when the liquid and vapour phases of soil-water are in "thermodynamic" equilibrium. Substituting $\psi_m = \psi_h$ in equation (C.6a) and inverting the logarithm we find the relative humidity of soil-air h as the "thermodynamic function" of soil-water potential ψ_m at any z (Edlefsen and Anderson, 1943; Penman, 1958; Rose, 1966; Campbell, 1977)

$$h(z) = \exp \frac{M_w \psi_m(z)}{RT} \tag{C.6b}$$

Given $\psi_m(z)$, relationship (C.6b) instantaneously partitions $\theta(\psi_m)$ between the liquid and vapour phases, and allows us to use one, rather than two, state variables for soil-water content. We have ignored the complication of hysteresis in this brief survey.

In the language of the thermodynamics of Gibbs, ψ is the partial specific free energy of the water in the soil-air-water system (Edlefsen and Anderson, 1943). The work performed by the process in converting water from state S_0 to S must be reversible, independent of path, and depend solely on the initial and final states. Necessary conditions for this, are usually given as

(a) the process is sufficiently slow to be represented as a succession of steady states,
(b) friction is negligible, and
(c) constant temperature (Rose, 1966).

Condition (a) contradicts equation (C.5), since this parabolic equation has the property that infinitesimal changes in ψ_m are propagated everywhere with infinite speed.

For a penetrating critique of the problematic foundations of classical thermodynamics and its development on a rigourous foundation, see Truesdell (1981), Truesdell et al. (1984) and Serrin et al. (1986).

C.4 LIQUID AND VAPOUR CONDUCTIVITIES

The conductivities for liquid and water vapour in soil at the pedon scale are k_l and k_v ($kg\,m^{-2}s^{-1}$). They depend on the properties of the soil matrix and on water content θ. The inverse function $\theta = \theta(\psi_m)$ of the soil-water characteristic may be used to express $k_l[\theta(\psi_m)]$ and $k_v[\theta(\psi_m)]$ as functions of ψ_m.

The liquid-water conductivity k_l is zero at zero water content, and is assumed to increase monotonically to a maximum at saturation. k_l is determined by measurement on soil samples at the pedon scale. In contrast to k_l, the vapour conductivity k_v is zero for both saturated and completely dry soil, and may rise to a maximum at a potential of roughly $-10^4\,J\,kg^{-1}$ (equivalent to 100,000 cm of water, $pF = 5$, when potential is defined as energy per unit weight). In the early stages of drying, the relative humidity of soil air is one and k_v is negligible compared with k_l. However, when the water potential in the soil drops below $-10^4\,J\,kg^{-1}$, mass transfer takes place almost entirely in the vapour phase.

The flux density of water vapour is given by the isothermal theory of vapour diffusion (Fick's law) in porous media (Penman, 1940; Philip, 1955; Jackson, 1964; Campbell, 1985)

$$f_v = -D_v \frac{\partial c}{\partial z} \tag{C.7a}$$

where c is the water vapour concentration in soil air ($kg\,m^{-3}$) and D_v the corresponding vapour diffusivity ($m^2\,s^{-1}$). D_v is given by the expression

$$D_v = -\alpha D_0[\theta_s - \theta(\psi_m)] \tag{C.7b}$$

where D_0 is the binary diffusion coefficient of water vapour in still air, at the prescribed temperature; $[\theta_s - \theta(\psi_m)]$ is the air-filled porosity of tortuosity α, through which vapour diffusion can occur.

Accordingly, using the chain rule for the derivative $\partial c/\partial z$, the conductivity k_v is

$$k_v = -\alpha D_0(\theta_s - \theta)c_{sat} \frac{dh}{d\psi_m} \tag{C.7c}$$

where $h = c/c_{sat}$ is the relative humidity and c_{sat} is the value of c at saturation for the given temperature T. From expression (C.6b) we can find $dh/d\psi_m$. Hence the vapour conductivity k_v can be expressed entirely in terms of ψ_m.

C.5 BOUNDARY CONDITIONS AT THE PEDON SCALE

Boundary conditions at the bottom of the soil column

The boundary condition at the bottom ($z = z_b$) of the column of unsaturated soil can be expressed in terms of the air-entry potential ψ_e,

$$\psi = \psi_e \quad \text{at } z = z_b \tag{C.8a}$$

$\psi_e < 0$ is the potential at which air first enters a saturated ($\theta = \theta_s$) sample of soil during drying, coinciding with the emptying of the largest pores in the soil matrix. This condition fixes the top of the saturated capillary fringe at a depth z_b below the surface. We will refer to z_b as the depth to the water table.

Water fluxes in either direction are possible across the lower boundary at z_b and represent capillary rise from or drainage to the saturated zone in the lower deposits of soil. A fixed value of z_b is adequate for simulating periods of infiltration or evaporation of a few days duration.

On longer time scales, z_b varies with time. z_b increases due to outflow of water to drains, and it decreases when rain-water infiltrates down to z_b. During prolonged periods of high evaporation z_b may also decrease.

We may derive a differential equation for $z_b(t)$ from a dynamic water balance on a column of saturated soil lying between z_b and z_d, the water level in the nearest surface drain.

The volume of water in a column of porosity ϕ_f per unit cross sectional area is $\phi_f(z_d - z_b)$. Assuming that the flow rate to (and from) the drain is proportional to the volume of water in the saturated column, we write the dynamic water balance

$$\rho_w \phi_f \frac{dz_b}{dt} = \kappa \rho_w \phi_f (z_d - z_b) - f_l(z_b) \tag{C.8b}$$

as where $\kappa(s^{-1})$ is a fractional rate of adjustment in the relative water level $(z_d - z_b)$ due to drainage. $f_l(z_b)$ is the flux density of liquid water (C.1b) at z_b. We may ignore the vapour flux contribution in this case. Equation (C.8b) is a first-order inhomogeneous ordinary differential equation for $z_b(t)$. It is also linear when the porosity ϕ_f is independent of depth. $f_l(z_b)$ couples equations (C.5) and (C.8b).

Equation (C.5) must now be solved subject to a moving boundary condition (C.8b). This is considerably more difficult than the fixed condition in (C.8a). The parameter κ may be found using field measurements of $z_b(t)$

or simple models of saturated flow to drains. A more detailed treatment requires a two-dimensional theory of combined saturated and unsaturated flow.

Surface boundary condition for infiltration of rain

The boundary condition for infiltration of rain into soil is a prescribed flux at $z = 0$, equal to the rainfall rate $Q\,(\mathrm{m^3\,m^{-2}\,s^{-1}})$

$$f_l = -k_l \frac{\partial \psi}{\partial z} = \rho_w Q \qquad \theta < \theta_s \quad \text{at } z = 0 \qquad (C.9a)$$

Vapour flux is not normally considered during infiltration.

As the solution of equations (C.5), (C.8a) and (C.9a) evolves in time, from a given initial condition $\theta(z, 0)$, $\theta(0, t)$ increases and may reach saturation. When saturation occurs at the ponding time t_p, the *first stage of infiltration* at the *atmosphere-controlled* rate is completed and the boundary condition *switches* to a prescribed concentration $\theta = \theta_s$

$$\theta(0, t_p) = \theta_s \qquad (C.9b)$$

As the solution of equations (C.5), (C.8a) and (C.9b) evolves in time, the flux into the soil $f_l(0, t)$ decreases from its initial value of $\rho_w Q$. During the second stage of ponded infiltration, the surplus water accumulates on the surface of the ground and starts to flow horizontally.

In the first stage of infiltration, the flux into the soil is atmosphere-controlled. In the second stage, it is soil-controlled. Analytical solutions for both linear and nonlinear simplifications of the above equations are summarised in Kuehnel (1989). Approximate expressions for the "time to ponding" for soils that are initially dry, are given in Kuehnel et al. (1990a). Time to ponding may vary by several orders of magnitude depending on soil properties and rainfall. If the rainfall intensity is less than the liquid conductivity at saturation, the switch to the second stage (ponding) does not occur. We now consider the soil after rainfall has stopped. The patches of ponded water, if there are any, will gradually disappear due to infiltration and evaporation. The infiltrated water will also be redistributed within the soil due to matric forces and gravity.

Surface boundary conditions for isothermal evaporation

In order to simplify the presentation, we begin with a column of bare soil with an initial water content but no ponded water on the surface. The initial water content is the accumulated effect of past rainfall, evaporation and drainage, modified by spatially varying topography, atmospheric and soil properties. The relative humidity of the air in the soil surface is 100% and evaporation begins at the potential rate E_p.

The boundary condition for evaporation at the surface is a flux E of water vapour from soil to atmosphere

$$f_l = -k_l \frac{\partial \psi}{\partial z} = 0 \qquad \text{at } z = 0 \tag{C.10a}$$

$$f_v = -k_v \frac{\partial \psi}{\partial z} = -\rho_w E \quad \text{at } z = 0 \tag{C.10b}$$

The actual evaporation rate E ($\text{m}^3 \, \text{m}^{-2} \, \text{s}^{-1}$) may be defined as follows (Brutsaert, 1982)

$$E = H(w)[c_0 - c_a] \tag{C.11}$$

where $H(w)$ is a mass transfer coefficient that varies with wind speed w at a reference height above the ground, c_0 is the water vapour concentration in the air in the soil surface and c_a is the water vapour concentration at the reference height in the atmosphere. In order to combine this boundary condition with the Philip–Richards equation, we must relate c_0 to ψ_m at $z = 0$.

The simplest relationship between c_0 and $\psi_m(0)$ is obtained by assuming an isothermal soil-atmosphere continuum. Hence, we may define a potential or maximum evaporation rate E_p ($\text{m}^3 \, \text{m}^{-2} \, \text{s}^{-1}$) as the evaporation rate from a wet soil at the same uniform temperature as the atmosphere

$$E_p = H(w)[c_{sat} - c_a] \tag{C.12}$$

where c_{sat} is the saturation water vapour concentration in the soil surface at the given temperature.

Numerical experiments by van Bavel and Hillel (1976), based on an albedo that varies with water content, show that potential evaporation is not a precise concept even under isothermal conditions. In general, evaporation depends on both the soil and the atmosphere. Nevertheless, it is a very useful first approximation.

Eliminating $H(w)$ from (C.11) and (C.12) we find (Philip, 1957a)

$$E = E_p \left[\frac{h_0 - h_a}{1 - h_a} \right] \qquad 1 \geq h_0 \geq h_a \tag{C.13}$$

where $h_a = c_a/c_{sat}$ is the relative humidity of the atmosphere at the reference height; $h_0 = c_0/c_{sat}$ is the relative humidity of the air in the soil surface at $z = 0$, and is found by inserting $\psi_m(0)$ into (C.6b).

During the isothermal drying of wet soils, the precise value of h_a is not important in equation (C.13) for small values of t, since $h_0 = 1$, $E/E_p = 1$ and E_p drives the model. Eventually h_0 may decrease from unity at a rate that depends on soil hydraulic properties, which control the upward supply of water. The evaporating surface, which is initially located in the soil surface at $z = 0$, starts to move downwards and transport of soil-water in the vapour phase appears. Eventually thermal gradients become important and must be considered.

Values of E_p may be estimated from Penman's (1948) equation using standard meteorological data or from open water evaporation rates after correction by a pan factor. A diurnal cycle may be introduced into E_p in proportion to the elevation of the sun (van Bavel and Hillel, 1976; Hillel, 1977).

Combining equations (C.10b), (C.13) and (C.14), we find a non-linear inhomogeneous boundary condition of the mixed type at $z = 0$

$$k_v \frac{\partial \psi_m}{\partial z} - \frac{\rho_w E_p}{1 - h_a} \exp\left[\frac{M_w \psi_m}{RT}\right] = k_v g - \frac{\rho_w E_p h_a}{1 - h_a} \tag{C.14}$$

for isothermal evaporation from wet bare soil at the scale of one meter.

At very low moisture content, thermal gradients cannot be neglected. An additional partial differential equation is required to model the flow of heat. The pair of partial differential equations for heat and moisture is coupled through the energy balance at the surface and also through their coefficients, in accordance with the theory of Philip and de Vries (1957). For numerical techniques for solving various forms of the above equations and their boundary conditions see Hillel (1977), Vauclin et al. (1979), Campbell (1985), Passerat de Silans (1986), and Scotter et al. (1988).

C.6 NON-LINEAR SWITCHING

The effect of the boundary condition (C.14) is to switch suddenly at some point in time from an evaporative flux controlled by the atmosphere to a flux controlled by the soil (Philip, 1957b, 1967, 1975; Rose, 1963; Jackson, 1964; Sasamori, 1970). The time of switching emerges from the combined solution of the Philip–Richards equation and the boundary condition (C.14). It is not prescribed in advance. This non-linear switching is mathematically similar to the switching during rain before and after ponding.

A simple two-stage approximation to (C.14) is sometimes used. In the first stage, a flux boundary condition equal to E_p is specified and lasts until the water content of the surface reaches a designated low value — the end of atmosphere control. A switch is then made to this low water content as the "concentration" boundary condition for the second stage — the start of soil control. The moisture profile at the end of the first stage provides the initial condition for the second stage. See Kuehnel et al. (1990b) for approximate analytical solutions.

Soil water responds to alternating wet and dry periods, each of which starts with a surface flux in or out of the soil that is set by atmospheric conditions. The moisture profile at the end of one period provides the initial condition for the next period. Within each period, a switch from atmosphere control to soil control may occur if the atmospheric conditions are strong enough and if they last for sufficient time. We refer to these times as the "time to ponding" in the case of infiltration of rain, and the "time to stage-two drying" in the case of evaporation. Consequently, there are two sets of switches in our computational clock: an outer pair always switching between wet and dry periods, and an inner pair switching intermittently from atmosphere control to soil control. This strongly non-linear behaviour is modelled by switching to and from the appropriate boundary conditions given in the above discussion. Simultaneously, the bottom boundary condition continues to play its role in contributing to the dynamic redistribution of water within the unsaturated soil column.

The action of the two pairs of switches is expressed spatially as a set of dynamic patches covering the soil surface, with each patch in one of four possible states: surface water flux in or out, surface water flux controlled by the atmosphere or by the soil. These patches have a life cycle. They grow, coalesce, shrink and disappear. Their shape may be fractal and will reflect the structure of the atmospheric boundary layer, the detailed geometry of the ground surface, the properties of the soil underneath and the deeper drainage in the saturated zone. They await discovery in remotely sensed data.

The transpiration of plants can be included as a distributed sink-term in the water balance equation. Plants extract water from the three-dimensional space occupied by their roots. In contrast, surface evaporation passes through a thin two-dimensional vapour layer, or bottleneck, at the surface of the soil. The presence of macro-pores, arising from biological processes, and also cracks, allows evaporation to occur from the three-dimensional volume of the soil. They also allow rapid infiltration to occur during rain. The above model must be modified when these factors are important. However, the modified model will retain the non-linear switches, which are seen most clearly in the case of bare homogeneous soil at the pedon scale. At larger scales, spatial averaging can be expected to attenuate their abrupt action.

For further detail, the reader is referred to the book by Slatyer (1967), the papers by Philip (1966, 1969) and more recent surveys of SVAT (Soil–vegetation–atmosphere) models at various scales.

REFERENCES

Brutsaert, W. (1982). *Evaporation into the Atmosphere.* D. Reidel, Dordrecht, 299 pp.

Buckingham, E. (1907). Studies on the movement of soil moisture. *U.S. Dep. Agric. Bur. Soils Bull.*, 38, pp. 28–31.

Campbell, G.S. (1977). *An Introduction to Environmental Biophysics.* Springer Verlag, New York, N.Y.

Campbell, G.S. (1985). *Soil Physics with BASIC — Transport Models for Soil–Plant Systems*. Elsevier, Amsterdam, 150 pp.

Edlefsen, N.E. and Anderson, A.B.C. (1943). The thermodynamics of soil moisture. *Hilgardia*, 16, pp. 31–299.

Feddes, R. (1971). *Water, Heat and Crop Growth*. Meded. Landbouwhogeschool Wageningen 71-12. The Institute for Land and Water Management, Wageningen.

Hillel, D. (1977). *Computer Simulation of Soil-Water Dynamics; A Compendium of Recent Work*. International Development Research Centre, Ottawa, p. 214.

Jackson, R.D. (1964). Water vapour diffusion in relatively dry soil: 1. Theoretical considerations and sorption experiments. *Soil Sci. Soc. Am. Pro.*, 28, pp. 172–176.

Kuehnel, V. (1989). *Scale Problems in Soil Moisture Flow*. Ph.D. Thesis, National University of Ireland.

Kuehnel, V., Dooge, J.C.I., Sander, G.C. and O'Kane, J.P. (1990a). Duration of atmosphere-controlled and of soil-controlled phases of infiltration for constant rainfall at a soil surface. *Ann. Geophys.*, 8, pp. 11–20.

Kuehnel, V., Dooge, J.C.I., Sander, G.C. and O'Kane, J.P. (1990b). Duration of atmosphere-controlled and of soil-controlled phases of evaporation for constant potential evaporation at a soil surface. *Ann. Geophys.*, 8, pp. 21–28.

Passerat de Silans, A. (1986). *Transferts de Masse et de Chaleur dans un Sol Stratifié Soumis a une Excitation Atmospherique Naturelle. Comparison: Modeles–Experience*. These de Docteur, l'Institut National Polytechnique de Grenoble.

Penman, H.L. (1940). Gas and vapour movement in soil. I and II. *J. Agr. Sci.*, 30, pp. 437–462, 570–581.

Penman, H.L. (1948). Natural evaporation from open water, bare soil and grass. *Proc. R. Soc. London* A, 193, pp. 120–145.

Penman, H.L. (1958). *Humidity*. Monograph of the Institute of Physics. Chapman and Hall, London.

Philip, J.R. (1955). The concept of diffusion applied to soil water. *Proc. Natl. A. Sci. India A*, 24, pp. 93–104.

Philip, J.R. (1957a). *The Physical Properties of Soil Water Movement during the Irrigation Cycle*. Proc. 3rd. Int. Congr. Int. Comm. on Irrigation and Drainage, pp. 8.125–8.154.

Philip, J.R. (1957b). Evaporation, and moisture and heat fields in the soil. *J. Meteorol.*, 14, pp. 354–366.

Philip, J.R. (1966). Plant water relations: some physical aspects. *Annu. Rev. Plant Phys.*, 17, pp. 245–268.

Philip, J.R. (1967). The second stage of drying of soil. *J. Appl. Meteorol.*, 6, pp. 581–582.

Philip, J.R. (1969). The soil–plant–atmosphere continuum in the hydro-logical cycle. In: *Hydrological Forecasting*, Tech. Note No. 92(1969), W.M.O. No. 228, TP No. 122, pp. 5–12.

Philip, J.R. (1975). Water movement in soil. In: D.A. de Vries and N.H. Afgan (eds.), Heat and Mass Transfer in the Biosphere. Papers from a Forum on Transfer Processes in the Plant Environment, held in Dubrovnik, August 1974. *Adv. Therm. Eng.*, 3, pp. 29–47.

Philip, J.R. and de Vries, D.A. (1957). Moisture movement in porour materials under temperature gradients. *Trans. Am. Geophys. Union*, 38, pp. 222–232.

Richards, L.A. (1931). Capillary conduction of liquids through porous mediums. *J. Phys.*, 1, pp. 318–333.

Rose, C.W. (1966). *Agricultural Physics*. Pergamon Press, London.

Rose, D.A. (1963). Water movement in porous materials. Part I — Isothermal vapour transfer. *Br. J. Appl. Phys.*, 14, pp. 256–262.

Sasamori, T. (1970). A numerical study of atmospheric and soil boundary layers. *J. Atmos. Sci.*, 27, pp. 1122–1137.

Scotter, D.R., Clothier, B.E. and Sauer, T.J. (1988). A critical assessment of the role of measured hydraulic properties in the simulation of absorp-tion, infiltration and redistribution of soil water. *Agric. Water Manage.*, 15, pp. 73–86.

Serrin, J. et al. (1986). *New Perspectives in Thermodynamics*. Springer-Verlag, Berlin.

Slatyer, R.O. (1967). *Plant–water relationships*. Experimental Botany, an Intern. Series of Monographs, Vol. II. London and New York, Academic Press, 366 pp.

Truesdell, C. (1981). *Textbook of Classical Thermodynamics*. Department of Mechanics, The Johns Hopkins University.

Truesdell, C. et al. (1984). *Rational Thermodynamics*, 2nd edition. Springer-Verlag, Berlin.

Van Bavel, C.H.M. and Hillel, D.I. (1976). Calculating potential and actual evaporation from a bare soil surface by simulation of concurrent flow of water and heat. *Agric. Meteorol.*, 17, pp. 453–476.

Vauclin, M., Haverkamp, R. and Vachaud, G. (1979). *Resolution Numerique d'une Equation de Diffusion non Lineaire. Application a l'Infiltration de l'Eau dans les Sols non Saturés*. Presses Universitaires de Grenoble, 183 pp.

APPENDIX D

Unsteady Flow in Open Channels

This appendix provides the background for the discussion of unsteady flow in open channels in Chapter 8. The question of non-linearity is the main theme of the presentation. The linearised channel response is derived and compared to several two and three parameter conceptual models of flood waves.

D.1 BASIC EQUATIONS

The continuity equation at a point in an incompressible fluid formulated by d'Alembert in 1744 was extended a hundred years later to a rectangular channel by Dupuit in 1848 and by Kleitz in 1858. They expressed it in the form

$$\frac{\partial(u \cdot y)}{\partial x} + \frac{\partial y}{\partial t} = 0 \tag{D.1}$$

where $y(x, t)$ is the depth of flow and $u(x, t)$ the average velocity of flow at position x and time t (Dooge, 1985). This was extended to any shape of section by St. Venant in 1871 who wrote

$$\frac{\partial(u \cdot A)}{\partial x} + \frac{\partial A}{\partial t} = 0 \tag{D.2}$$

where $A(x, t)$ is the area of flow. $A(x, t)$ may also be interpreted as the volume of water per unit length at any point on a curvilinear axis x.

The lumped form of the continuity equation for a channel reach

$$Q_A(t) - Q_B(t) = \frac{dS_{AB}}{dt} \tag{D.3}$$

where S_{AB} is the amount of water stored between the inflow section A and the outflow section B, was applied to lakes by William Mulvany in 1851 and by Graeff in 1875. It was later applied to channel reaches but the date of its first use for this purpose is not clear.

Kleitz appears to have derived the momentum equation corresponding to equation (D.3) but his memorandum circulated in 1858 was never published. The classical formulation of the momentum equation is that due to St. Venant who wrote in 1871:

$$\frac{\partial z}{\partial x} + \frac{1}{g}\frac{\partial u}{\partial t} + \frac{u}{g}\frac{\partial u}{\partial x} + \tau_0\frac{P}{A} = 0 \tag{D.4}$$

where $z(x, t)$ is the elevation of water surface, $\tau_0(x, t)$ the average boundary shear, and $P(x, t)$ the wetted perimeter. For uniform bed slope and cross section with Chezy friction, this equation is currently used in the form

$$\frac{\partial y}{\partial x} + \frac{u}{g}\frac{\partial u}{\partial x} + \frac{1}{g}\frac{\partial u}{\partial t} = S_0 - S_f \tag{D.5}$$

where S_0 is the downward slope of the channel and S_f the friction slope.

If there is a lateral inflow, then the equation of continuity given by (D.2) must be adjusted as follows:

$$\frac{\partial Q}{\partial x} + \frac{\partial A}{\partial t} = r(x, t) \tag{D.6}$$

where $r(x, t)$ is the rate of lateral inflow of water volume per unit time per unit length of x. Similarly equation (D.5) must be adjusted to allow for the momentum drag on the main flow by the lateral inflow and is written as

$$\frac{\partial y}{\partial x} + \frac{u}{g}\frac{\partial u}{\partial x} + \frac{1}{g}\frac{\partial u}{\partial t} = S_0 - S_f - \frac{u}{gy}r(x, t) \tag{D.7}$$

Equations (D.6) and (D.7) appear in the main text on page 163 as equations (8.1) and (8.2) respectively.

The main features of these equations, which create problems for their solution, can be illustrated for the special case of a wide rectangular channel with Chezy friction and a constant rate of lateral inflow r_0. When this is done, the equation of continuity given by equation (D.6) is reduced to

$$\frac{\partial q}{\partial x} + \frac{\partial y}{\partial t} = r_0 \tag{D.8}$$

which appears in the main text on page 164 as equation (8.3) and the momentum equation given by equation (D.7) reduces to

$$\frac{\partial y}{\partial x} + \frac{u}{g}\frac{\partial u}{\partial x} + \frac{1}{g}\frac{\partial u}{\partial t} = S_0 - \frac{u^2}{C^2 y} - \frac{u}{gy}r_0 \tag{D.9}$$

which appears in the main text on page 164 as equation (8.4a).

Reduction to a single dependent variable

Equations (D.8) and (D.9) contain three dependent variables: $q(x, t), y(x, t)$ and $u(x, t)$. Since these three variables are linked by the definition equation

$$q(x, t) = u(x, t)\,y(x, t) \tag{D.10}$$

equations (D.8) and (D.9) can be transformed to two first order differential equations in terms of any two of these variables. Since equation (D.8) is simple and linear, it is advisable to maintain it in its original form and replace $u(x, t)$ by $q(x, t)/y(x, t)$. When this is done (D.9) becomes

$$\frac{\partial y}{\partial x} + \frac{q}{gy}\left(y\frac{\partial q}{\partial x} - q\frac{\partial y}{\partial x}\right)\frac{1}{y^2} + \frac{1}{g}\left(y\frac{\partial q}{\partial t} - q\frac{\partial y}{\partial t}\right)\frac{1}{y^2}$$

$$= S_0 - \frac{q^2}{C^2 y^3} - \frac{q}{gy^2}r_0 \tag{D.11a}$$

Multiplying across by (gy^3) gives

$$gy^3\frac{\partial y}{\partial x} + q\left(y\frac{\partial q}{\partial x} - q\frac{\partial y}{\partial x}\right) + y\left(y\frac{\partial q}{\partial t} - q\frac{\partial y}{\partial t}\right)$$

$$= gy^3 S_0 - \frac{g}{C^2}q^2 - yqr_0 \tag{D.11b}$$

Now (qy) times equation (D.8) is

$$qy\frac{\partial q}{\partial x} + qy\frac{\partial y}{\partial t} = qyr_0 \tag{D.11c}$$

Adding equations (D.11b and c) cancels the terms containing the lateral inflow and yields

$$\left(gy^3 - q^2\right)\frac{\partial y}{\partial x} + 2qy\frac{\partial q}{\partial x} + y^2\frac{\partial q}{\partial t} = S_0 gy^3 - \frac{g}{C^2}q^2 \tag{D.11d}$$

which appears in the main text on page 164 as equation (8.4b) and is repeated on page 180. It is non-linear in all five terms.

Alternatively equation (D.9) which is in terms of $y(x, t)$ and $u(x, t)$ can be left intact and equation (D.8) written as

$$\frac{\partial y}{\partial x} + \frac{\partial u}{\partial x} + \frac{\partial y}{\partial t} = r_0 \tag{D.12}$$

which also remains linear. In either case we have two first order differential equations in two variables. Standard numerical techniques can be applied to the solution of either formulation of the problem but analytical solutions are not available because of the inherent non-linearity of the momentum equation.

By making use of differentiation, these two first order equations can be combined into a single second order differential equation in either $q(x, t)$ or $y(x, t)$; but this equation is found to be more complex in form than either of the original equations (D.8) and (D.9). Since the next step in seeking a solution is to linearise the resulting equations, this complexity can be avoided by linearising before reducing to a single variable. The final linearised equation in a single variable is the same whichever route is taken.

Equation (D.8) is already linear in form. The momentum equation can be linearised by expressing the variables q and y as perturbations $q'(x, t)$ and $y'(x, t)$ from a fixed reference condition (q_0, y_0)

$$q(x, t) = q_0(x) + q'(x, t) \tag{D.13a}$$

and

$$y(x, t) = y_0(x) + y'(x, t) \tag{D.13b}$$

where the reference values satisfy

$$\frac{\partial q_0}{\partial x} + \frac{\partial y_0}{\partial t} = r_0 \tag{D.13c}$$

Consequently the perturbations satisfy a continuity equation of the same general form, i.e.

$$\frac{\partial q'}{\partial x} + \frac{\partial y'}{\partial t} = 0 \tag{D.13d}$$

Equation (D.11d) can then be linearised by freezing the values of the coefficients of the derivative terms on the basis of the same reference condition (q_0, y_0) by choosing a value of q_0 or y_0 and calculating the other reference value from the *steady state rating curve*

$$\frac{\partial y_0}{\partial t} = 0 \tag{D.14a}$$

$$q_0 = C y_0 \sqrt{S_0 y_0} \tag{D.14b}$$

and

$$\frac{\partial q_0}{\partial x} = \frac{d q_0}{d x} = r_0 \tag{D.14c}$$

which is equivalent to solving for the first order perturbations from the chosen steady-state reference condition.

When this approach is applied to equation (D.11), the linearised equation in terms of the perturbations $q'(x, t)$ and $y'(x, t)$ is given by

$$\left(g y_0^3 - q_0^2 \right) \frac{\partial y'}{\partial x} + 2 q_0 y_0 \frac{\partial q'}{\partial x} + y_0^2 \frac{\partial q'}{\partial t} = S_0 g y^3 - \frac{g}{C^2} q^2 - q y r_0 \tag{D.15}$$

The perturbations enter all the derivatives on the left-hand side. In contrast, on the right-hand side, the original variables remain in all three terms; the lateral inflow r is multiplied by the unperturbed qy, the multiplier of the continuity equation (D.11c). For further analysis it is convenient to combine equations (D.14) and (D.15) into one second order equation involving only $q'(x, t)$, or only $y'(x, t)$.

The first step in the reduction of the two equations to a single equation in terms of $q'(x, t)$ is to differentiate equation (D.15) with respect to time and freeze the coefficients of new derivative terms on the right-hand side at the reference condition. The result is

$$
\left(gy_0^3 - q_0^2\right)\frac{\partial^2 y'}{\partial x \partial t} + 2q_0 y_0 \frac{\partial^2 q'}{\partial x \partial t} + y_0^2 \frac{\partial^2 q'}{\partial t^2}
$$

$$
= 3gS_0 y_0^2 \frac{\partial y'}{\partial t} - \frac{2gq_0}{C^2}\frac{\partial q'}{\partial t} - q_0 r_0 \frac{\partial y'}{\partial t} - y_0 r_0 \frac{\partial q'}{\partial t} \tag{D.16}
$$

The only term involving $y'(x,t)$ on the left hand side of the above equation can be expressed in terms of $q'(x,t)$ by differentiating the continuity equation (D.14) with respect to x to obtain

$$
\frac{\partial^2 q'}{\partial x^2} + \frac{\partial^2 y'}{\partial x \partial t} = 0 \tag{D.17}
$$

The terms involving $y'(x,t)$ on the right hand side of equation (D.16) can be expressed in terms of $q'(x,t)$ by the direct application of equation (D.14) in its original form. When these substitutions are made, the resulting equation is

$$
\left(gy_0^3 - q_0^2\right)\frac{\partial^2 q'}{\partial x^2} - 2q_0 y_0 \frac{\partial^2 q'}{\partial x \partial t} - y_0^2 \frac{\partial^2 q'}{\partial t^2}
$$

$$
= \left(3gS_0 y_0^2 - q_0 r_0\right)\frac{\partial q'}{\partial x} + \left(\frac{2gq_0}{C^2} + y_0 r_0\right)\frac{\partial q'}{\partial t} \tag{D.18a}
$$

Dividing across by y_0^2 yields

$$
\left(1 - \frac{q_0^2}{gy_0^3}\right)gy_0 \frac{\partial^2 q'}{\partial x^2} - 2\frac{q_0}{y_0}\frac{\partial^2 q'}{\partial x \partial t} - \frac{\partial^2 q'}{\partial t^2}
$$

$$
= \left(3gS_0 - \frac{q_0 r_0}{y_0^2}\right)\frac{\partial q'}{\partial x} + \left(\frac{2gq_0}{C^2 y_0^2} + \frac{r_0}{y_0}\right)\frac{\partial q'}{\partial t} \tag{D.18b}
$$

Inserting the Froude number $F_0^2 = q_0^2/gy_0^3$ and the steady rating-curve relationship (D.14b) $S_0 = q_0^2/C^2 y_0^3$ for the reference condition into (D.18b) yields

$$
\left(1 - F^2\right)gy_0 \frac{\partial^2 q'}{\partial x^2} - 2u_0 \frac{\partial^2 q'}{\partial x \partial t} - \frac{\partial^2 q'}{\partial t^2}
$$

$$
= \left(3gS_0 - \frac{u_0 r_0}{y_0}\right)\frac{\partial q'}{\partial x} + \left(\frac{2gS_0}{u_0} + \frac{r_0}{y_0}\right)\frac{\partial q'}{\partial t} \tag{D.18c}
$$

For the case of upstream inflow only (i.e. $r_0 = 0$) this is equation (8.36) on page 180 of the main text without the prime.

The derivation is similar for the case of $y'(x, t)$ as the single dependent variable, the only difference being that the continuity equation is differentiated with respect to time rather than distance giving

$$\frac{\partial^2 q'}{\partial x \partial t} + \frac{\partial^2 y'}{\partial t^2} = 0 \tag{D.19}$$

and the momentum equation is differentiated with respect to distance rather than time giving the following second order linearised differential equation for $y'(x, t)$:

$$\left(gy_0^3 - q_0^2\right) \frac{\partial^2 y'}{\partial x^2} - 2q_0 y_0 \frac{\partial^2 y'}{\partial x \partial t} - y_0^2 \frac{\partial^2 y'}{\partial t^2}$$

$$= \left(3gS_0 y_0^2 - q_0 r_0\right) \frac{\partial y'}{\partial x} + \left(\frac{2gq_0}{C^2} + y_0 r_0\right) \frac{\partial y'}{\partial t} \tag{D.20}$$

which is identical in form to equation (D.18a) for $q'(x, t)$.

D.2 LINEARISED CHANNEL RESPONSE

Since equations (D.18) and (D.20) are linear, it is only necessary to solve them for a delta-function input and then convolute this response function with the actual boundary condition. The solution for a semi-infinite channel with an upstream inflow ($r_0 = 0$) is termed the linear channel response, or the LCR (Dooge and Harley, 1965). The most convenient method of solution is the use of the Laplace transform.

If we denote the linear channel response as $h(x, t)$ then its Laplace transform is given by

$$H(x, s) = \int_0^\infty \exp(-st)\, h(x, t)\, dt \tag{D.21}$$

and equation (D.18a) or equation (D.20) for the case $r_0 = 0$ becomes in transform domain

$$\left(gy_0^3 - q_0^2\right) \frac{d^2 H}{dx^2} - 2q_0 y_0 s \frac{dH}{dx} - y_0^2 s^2 H$$

$$= \left(3gS_0 y_0^2\right) \frac{dH}{dx} + \left(\frac{2gq_0}{C^2}\right) sH \tag{D.22}$$

The solution of this second-order differential equation is given by

$$H(x, s) = A_1(s) \exp[\lambda_1(s)x] + A_2(s) \exp[\lambda_2(s)x] \tag{D.23}$$

where $A_1(s)$ and $A_2(s)$ are determined by the transformed boundary conditions and λ_1 and λ_2 are the roots of corresponding characteristic equation

$$\left(gy_0^3 - q_0^2\right) \lambda^2 - \left(2q_0 y_0 s + 3gS_0 y_0^2\right) \lambda - \left(y_0^2 s^2 + \frac{2gq_0}{C^2} s\right) = 0 \tag{D.24a}$$

Rearranging the terms of the equation for solution we find

$$\lambda^2 - \left[\frac{2q_0 y_0}{(gy_0^3 - q_0^2)} s + \frac{3g S_0 y_0^2}{(gy_0^3 - q_0^2)} \right] \lambda$$

$$- \left[\frac{y_0^2}{(gy_0^3 - q_0^2)} s^2 + \frac{2g q_0}{C^2 (gy_0^3 - q_0^2)} s \right] = 0 \qquad \text{(D.24b)}$$

$$\lambda^2 - \left[\frac{2u_0}{(1 - F_0^2)gy_0} s + \frac{3S_0}{(1 - F_0^2)y_0} \right] \lambda$$

$$- \left[\frac{1}{(1 - F_0^2)gy_0} s^2 + \frac{2S_0}{(1 - F_0^2)u_0 y_0} s \right] = 0 \qquad \text{(D.24c)}$$

In other words

$$\lambda^2 + B\lambda + C = 0 \qquad \text{(D.24d)}$$

where

$$B = -\left[\frac{2u_0}{(1 - F_0^2)gy_0} s + \frac{3S_0}{(1 - F_0^2)y_0} \right]$$

$$C = -\left[\frac{1}{(1 - F_0^2)gy_0} s^2 + \frac{2S_0}{(1 - F_0^2)u_0 y_0} s \right] \qquad \text{(D.24e)}$$

The solution of this quadratic equation in λ is

$$\lambda = -\frac{B}{2} \pm \sqrt{\frac{B^2}{4} - C} \qquad \text{(D.25a)}$$

The first term on the right hand side is

$$-\frac{B}{2} = \frac{u_0}{gy_0} \frac{1}{(1 - F_0^2)} s + \frac{3}{2} \frac{S_0}{y_0} \frac{1}{(1 - F_0^2)} = es + f \qquad \text{(D.25b)}$$

which we square to find the second term

$$\frac{B^2}{4} - C = \left(\frac{2u_0}{(1 - F_0^2)gy_0} s + \frac{3S_0}{(1 - F_0^2)y_0} \right)^2 \Big/ 4 - C$$

$$= \frac{u_0^2}{(1 - F_0^2)^2 (gy_0)^2} s^2 + \frac{3u_0 S_0}{(1 - F_0^2)^2 gy_0^2} s + \frac{9 S_0^2}{4(1 - F_0^2)^2 y_0^2}$$

$$+ \frac{1}{(1 - F_0^2)gy_0} s^2 + \frac{2S_0}{(1 - F_0^2)u_0 y_0} s$$

$$= \frac{u_0^2 + gy_0(1 - F_0^2)}{(1 - F_0^2)^2(gy_0)^2}s^2 + \frac{3u_0^2S_0 + 2gy_0S_0(1 - F_0^2)}{(1 - F_0^2)^2u_0gy_0^2}s + \frac{9S_0^2}{4(1 - F_0^2)^2y_0^2}$$

$$= \frac{1}{gy_0}\frac{1}{(1 - F_0^2)^2}s^2 + \frac{S_0}{q_0}\frac{(2 + F_0^2)}{(1 - F_0^2)^2}s + \frac{9}{4}\left(\frac{S_0}{y_0}\right)^2\frac{1}{(1 - F_0^2)^2}$$

$$= as^2 + bs + c \qquad\qquad\qquad\qquad (D.25c)$$

The roots (λ_1, λ_2) are given by

$$\lambda = es + f \pm \sqrt{as^2 + bs + c} \qquad\qquad\qquad (D.25d)$$

where the positive root corresponds to upstream flow and the negative root to downstream inflow. The parameters (a, b, c, e, f) derived from the coefficients in equation (D.25b and c) depend on the properties of the channel at the reference conditions. For a delta function input at the upstream boundary of a semi-infinite channel, the solution is given by

$$H(x, s) = \exp(\lambda_2(s)x) \qquad\qquad\qquad (D.26a)$$

where

$$\lambda_2 = es + f - \sqrt{as^2 + bs + c} \qquad\qquad\qquad (D.26b)$$

which appears in the main text on page 181 as equation (8.37).

For the case of a wide rectangular channel with Chezy friction described by equation (D.24a), the values of the parameters are given by

$$a = \frac{1}{gy_0}\frac{1}{(1 - F_0^2)^2} \qquad\qquad\qquad (D.27a)$$

$$b = \frac{S_0}{q_0}\frac{(2 + F_0^2)}{(1 - F_0^2)^2} \qquad\qquad\qquad (D.27b)$$

$$c = \frac{9}{4}\left(\frac{S_0}{y_0}\right)^2\frac{1}{(1 - F_0^2)^2} \qquad\qquad\qquad (D.27c)$$

$$d = \frac{b^2}{4} - ac = \left(\frac{S_0}{q_0}\right)^2\frac{(1 - F_0^2/4)}{(1 - F_0^2)^3} \qquad\qquad\qquad (D.27d)$$

$$e = \frac{u_0}{gy_0}\frac{1}{(1 - F_0^2)} \qquad\qquad\qquad (D.27e)$$

$$f = \frac{3}{2}\frac{S_0}{y_0}\frac{1}{(1 - F_0^2)} \qquad\qquad\qquad (D.27f)$$

Since the logarithm of the Laplace transform can be used as a generating function of the cumulants of any function, the exponential form of equation (D.26a) simplifies the process of deriving these cumulants, the first three of which are given in equation (8.38) on page 181 of the main text.

The inversion of the linear channel response from the Laplace transform domain to the original time domain is possible but complicated. The solution consists of two parts:

(a) an initial wave which is in the form of a delta function and which travels downstream along the leading characteristic at the dynamic wave speed of $[u_0 + (gy_0)^{1/2}]$ but with an exponentially decreasing volume, and

(b) the body of the wave whose average speed decreases asymptotically towards the kinematic wave speed with increasing values of F_0. The form of the latter attenuated body of the wave is that of a modified Bessel function which is exponentially damped as it moves downstream.

The equations governing both parts of the linear channel response in the time domain are given on pages 182 and 183 of the main text. For the case of a wide rectangular channel with Chezy friction, the values of the r, s and h parameters in equation (8.49) are given by (Dooge and Harley, 1967)

$$r = \frac{2 + F_0^2}{F_0^2} \left(\frac{S_0 u_0}{2 y_0} \right) \tag{D.27g}$$

$$s = \frac{S_0}{2 y_0} \tag{D.27h}$$

$$h = \frac{\sqrt{\left(1 - F_0^2/4\right)\left(1 - F_0^2\right)}}{F_0^2} \left(\frac{S_0 u_0}{2 y_0} \right) \tag{D.27i}$$

For the more general case following equation (8.54) where the friction law takes the form

$$Q = k A^m \tag{D.28}$$

The value of m enters the expressions for the parameters defined by equations (8.48) and (8.51) but their form remains the same (Dooge, Napiorkowski and Strupczewski, 1987a).

D.3 APPROXIMATION BY TWO-PARAMETER CONCEPTUAL MODELS

In applied hydrology, conceptual models for flood routing have been widely used over the past sixty years. The three most widely used were the Muskingum Method proposed by McCarthy in 1939, the lag and route method proposed by Meyer in 1941, and the cascade of characteristic lengths proposed by Kalinin and Milyukov in 1957. All three models have a common point in the space of shape factors (defined in equation (5.3) in the text: p. 86) at $(s_2 = 1, s_3 = 2)$ which corresponds to the one parameter model of a single linear reservoir. This point represents the Muskingum model with the inflow-weighting parameter $x = 0$, the lag and route model with the lag parameter $T = 0$, and the Kalinin–Milyukov model with the number of reservoirs in the cascade n equal to one.

It is interesting to compare these linear conceptual models with the linearised solution for the St. Venant equations given above. This is most conveniently done by the use of shape factor diagrams applied to the comparison between conceptual models of catchment response on page 85 to 99 of Chapter 5 in the main text. The general form of the linearised equation of unsteady downstream flow in open channels is given by equation (8.54) on page 185 of the main text and the expression for the first three cumulants (i.e. the first moment about the origin, second moment about the centre, and the third moment about the center) are given by equation (8.55) on the same page.

For the lower limiting value of $F_0 = 0$ (i.e. the diffusion analogy), and for any value of m, where m in (D.28) is the ratio at the reference condition

$$m = \frac{\text{kinematic wave speed}}{\text{average velocity}} \tag{D.29}$$

the second and third order shape factors based on the first cumulant are given by

$$s_2 = \frac{k_2}{(k_1)^2} = \frac{1}{m} \frac{y_0}{S_0 x} \tag{D.30a}$$

$$s_3 = \frac{k_3}{(k_1)^3} = 3\frac{1}{m^2} \left(\frac{y_0}{S_0 x}\right)^2 \tag{D.30b}$$

So the relationship between them is:

$$s_3 = 3s_2^2 \tag{D.31}$$

and is independent of the value of the parameter m. For the upper limiting value of $F_0 = 1$ (i.e. critical flow) the corresponding limit is given by

$$s_3 = \frac{3}{2-m} (s_2)^2 \tag{D.32}$$

which is equal to the value for $F_0 = 0$ for the lower limit of $m = 1$ and equal to infinity for the upper limit of $m = 2$. For the case of the wide rectangle channel with Chezy friction ($m = 3/2$) the coefficient in equation (D.32) is equal to 6. For a wide rectangular channel with Manning friction ($m = 5/3$) the corresponding coefficient is 9. The limiting relationships of equation (D.31) and equation (D.32) can be plotted on a shape factor diagram and compared to the three conceptual models already mentioned (see Dooge, 1973, p. 258).

The shape factors for the single reach Muskingum model (McCarthy, 1938) are given by (see pages 31 and 32 and problem 10b, page 204)

$$s_2 = (1 - 2x) \tag{D.33a}$$

$$s_3 = 2(1 - 3x + 3x^2) \tag{D.33b}$$

which obey the relationship

$$s_3 = \frac{3s_2^2 + 1}{2} \tag{D.33c}$$

This method was used in applied hydrology in an empirical fashion. It was found as a rule of thumb that the value of the weighting factor x usually lay in the approximate range of $0.2 < x < 0.3$. If equation (D.33c) is compared with equations (D.31) and (D.32), we find that over a certain range equation (D.33c) lies within the region for the case of a wide rectangular channel with Chezy friction. This range is given by

$$\frac{1}{\sqrt{3}} < s_2 < \frac{1}{3} \tag{D.34a}$$

Since for the Muskingum model we have from equation (D.33a)

$$s_2 = (1 - 2x) \tag{D.33a}$$

this is equivalent to

$$0.21 < x < 0.33 \tag{D.34b}$$

which echoes the practice in applied hydrology.

For shorter channels i.e. for larger values of s_2 than the range in equation (D.34a), the value of s_3 predicted by the Muskingum model falls increasingly below the lower limit of the linearised solution. For longer channels s_3 falls increasingly above the upper limit and reaches an asymptotic value of $s_3 = 1/2$ for $s_2 = 0$ compared with the value of $s_3 = 0$ for the linearised solution.

In contrast, the lag and route method (Meyer, 1941) works best for longer channels. The first three cumulants are given by (page 94)

$$k_1 = K + T \tag{D.35a}$$
$$k_2 = K^2 \tag{D.35b}$$
$$k_3 = 2K^3 \tag{D.35c}$$

where T is the lag of the linear channel and K the lag of the linear reservoir.

The shape factors for the lag and route model based on the first cumulant are therefore given by

$$s_2 = \frac{K^2}{(K + T)^2} = \frac{1}{(1 + T/K)^2} \tag{D.36a}$$

and

$$s_3 = \frac{2K^3}{(K + T)^3} = \frac{2}{(1 + T/K)^3} \tag{D.36b}$$

The combination of these two equations gives us the relationship

$$s_3 = 2(s_2)^{3/2} \tag{D.37}$$

For values of s_2 less than 4/9, this curve falls within the region of the linearised solution for the complete equations. For larger values of s_2 (i.e. for shorter channel lengths) the value of s_3 is always less than the value for $F_0 = 0$ and for any value of m.

Since the Kalinin–Milyukov (1957) solution takes the form of a cascade of equal linear reservoirs, the linear channel response is given by a two-parameter gamma distribution of equation (5.19) on page 91 of the main text. The corresponding (s_3, s_2) relationship takes the form

$$s_3 = 2(s_2)^2 \tag{D.38}$$

which is the relationship given by equation (5.28) on page 92 of the main text. The value of s_3 given by equation (D.38) is always two-thirds of that for the lower limit of the linearised solution, and hence the Kalinin–Milyukov model always plots outside the region of the linearised solution of the complete St. Venant equations.

D.4 REPRESENTATION BY THREE-PARAMETER MODELS

It is interesting to analyse the improvement that can be achieved by adding a further parameter to the conceptual models discussed in the last section. This involves a more complex analysis. The initial discussion will be limited to the lag and multiple route model. The lag and route model is the special case of this model with $n = 1$ and the Kalinin–Milyukov model is the special case with $T = 0$. This three parameter model can be variously described as the lagged uniform linear cascade or the linear channel plus linear cascade model or the 3 parameter gamma distribution (Dooge, 1959). In the case of this and other lagged models, only the first cumulant is affected by the value of the lag parameter T. Consequently, the shape of the channel parameters of the response can be matched to second and third cumulants to determine the shape of the response function. Then the lag of the derived shape is subtracted from the observed lag to determine the appropriate lag of the linear channel as the third parameter.

Because the first cumulant of the derived shape is not equal to observed lag, the analysis cannot be conducted on the basis of the shape factors s_2 and s_3 defined by equation (5.3) on page 86 of the main text. Instead of basing the analysis on s_2 and s_3, we now define a new dimensionless shape factor f_R based on the second cumulant in the form

$$f_R = \frac{k_R}{(k_2)^{R/2}} \tag{D.39a}$$

and seek a relationship given by

$$f_4 = \phi(f_3) \tag{D.39b}$$

where

$$f_3 = \frac{k_3}{(k_2)^{3/2}} \tag{D.39c}$$

and

$$f_4 = \frac{k_4}{(k_2)^2} \tag{D.39d}$$

Since the fourth cumulant represents *excess kurtosis*, the accuracy of the prediction of the peak is strongly influenced by the accuracy of the estimate of k_4. Since the lag derived from the shape factor analysis is adjusted to fit the observed lag, the time to peak should be predicted to a high order of accuracy.

For any shape of section and friction law, the shape factors f_R defined by equation (D.39a) can be derived in terms of the parameters m and F_0 (Dooge, Napiorkowski and Strupczewski 1987b; Romanowicz, Dooge and Kundzewicz, 1988). For $F_0 = 0$ and any value of m, we have

$$f_3 = 3\sqrt{\frac{y_0}{mS_0x}} \tag{D.40a}$$

and

$$f_4 = 15\frac{y_0}{mS_0x} \tag{D.40b}$$

so that

$$f_4 = \frac{5}{3}f_3^2 \tag{D.40c}$$

Similarly for $F_0 = 1$ and for any value of m

$$f_3 = 3\sqrt{\frac{y_0}{(2-m)S_0x}} \tag{D.41a}$$

and

$$f_4 = 12\frac{y_0}{(2-m)S_0x} \tag{D.41b}$$

so that

$$f_4 = \frac{4}{3}f_3^2 \tag{D.41c}$$

for any value of m.

For the cascade of equal linear reservoirs we have

$$f_3 = \frac{2}{\sqrt{n}} \tag{D.42a}$$

$$f_4 = \frac{6}{n} \tag{D.42b}$$

so that

$$f_4 = \frac{3}{2}f_3^2 \tag{D.42c}$$

which lies midway between the two limits given by equations (D.40c) and (D.41c).

A similar approach using a linear channel plus a single Muskingum reach does not prove fruitful. Since for a single Muskingum reach the second moment about the centre, which is essentially positive, is given by

$$k_2 = (1 - 2x)K^2 \tag{D.43a}$$

and consequently to preserve this condition the weighting parameter x must be subject to the restriction

$$-\infty < x < \frac{1}{2} \tag{D.43b}$$

The third moment about the centre is given by

$$k_3 = 2(1 - 3x + 3x^2)K^3 \tag{D.44a}$$

And hence we can express the shape factor f_3 defined by equation (D.39c) in the form

$$f_3^2 = \frac{k_3^2}{k_2^3} = \frac{4\left(1 - 3x + 3x^2\right)^2}{(1 - 2x)^3} \tag{D.44b}$$

At the two limits given by equation (D.43b), the value of the right hand side of equation (D.44b) reaches infinity. The intermediate values are all positive with a minimum value of $f_3 = 2$ at $x = 0$ which corresponds to a single linear reservoir. Accordingly the model can only be applied to channel reaches shorter than the characteristic length.

Laurenson (1959) used an alternative approach in the form of a cascade of Muskingum reaches. For this model

$$k_1 = nK \tag{D.45a}$$

$$k_2 = (1 - 2x)nK^2 \tag{D.45b}$$

$$k_2 = 2(1 - 3x + 3x^2)nK^3 \tag{D.45c}$$

$$k_4 = 6(1 - 4x + 6x^2 - 4x^3)nK^4 \tag{D.45d}$$

The matching value of the weighting parameter x can be determined from the observed values of the first three moments (k_1, k_2, k_3) through the relationship

$$\frac{k_3 k_1}{k_2^2} = \frac{2\left(1 - 3x + 3x^2\right)}{(1 - 2x)^2} \tag{D.46a}$$

When this value has been obtained it can be used to derive an estimate of the lag of a single Muskingum element from

$$\frac{k_2}{k_1} = (1 - 2x)K \tag{D.46b}$$

and the number of such elements from

$$\frac{k_1^2}{k_2} = \frac{n}{(1 - 2x)} \tag{D.46c}$$

thus obtaining the complete set of parameters (x, k, n) for the three parameter model.

Equations (D.40c) and (D.41c) for the linearised channel response in tranquil flow (i.e. $0 < F_0 < 1$) indicate that for any value of the parameter m

$$\frac{4}{3}f_3^2 < f_4 < \frac{5}{3}f_3^2 \tag{D.47}$$

and equation (D.42c) indicates that the three parameter models composed of a linear channel and a cascade of equal linear reservoirs plots half way between these limits. The same comparison can be made for the multiple Muskingum model. Using the expression for k_2, k_3 and k_4 in equation (D.45) we find

$$\frac{f_4}{(f_3)^2} = \frac{k_4 k_2}{k_3^2} = \frac{3}{2} \frac{\left(1 - 4x + 6x^2 - 4x^3\right)(1 - 2x)}{(1 - 3x + 3x^2)^2} \tag{D.48a}$$

which has the same quadratic relationship between f_4 and f_3 as the other two cases.

For any realistic positive value of x (i.e. $0 < x < 0.5$) we have

$$0 < \frac{f_4}{f_3^2} < \frac{3}{2} \tag{D.48b}$$

and for any negative value of x $(-\infty < x < 0)$ we have

$$\frac{4}{3} < \frac{f_4}{f_3^2} < \frac{3}{2} \tag{D.48c}$$

The upper limiting value of $3/2$ corresponds to $x = 0$ i.e. the cascade of equal linear reservoirs. The model will fall within the area of the linear channel response, i.e. f_4 will exceed the limit of equation (D.41c), for the range of values of x given by

$$-\infty < x < 0.211 \tag{D.49}$$

The extension of the range of x to cover negative values enables us to extend the scope of the model beyond the limitation to short reaches imposed by restricting x to positive values.

Index